Reliability Analysis of Electrotechnical Devices

Reliability Analysis of Electrotechnical Devices

Editor

Cher Ming Tan

MDPI • Basel • Beijing • Wuhan • Barcelona • Belgrade • Manchester • Tokyo • Cluj • Tianjin

Editor
Cher Ming Tan
Chang Gung University
Taiwan

Editorial Office
MDPI
St. Alban-Anlage 66
4052 Basel, Switzerland

This is a reprint of articles from the Special Issue published online in the open access journal *Applied Sciences* (ISSN 2076-3417) (available at: https://www.mdpi.com/journal/applsci/special_issues/reliability_Analysis_Electrotechnical_Devices).

For citation purposes, cite each article independently as indicated on the article page online and as indicated below:

LastName, A.A.; LastName, B.B.; LastName, C.C. Article Title. *Journal Name* **Year**, *Volume Number*, Page Range.

ISBN 978-3-0365-4653-7 (Hbk)
ISBN 978-3-0365-4654-4 (PDF)

© 2022 by the authors. Articles in this book are Open Access and distributed under the Creative Commons Attribution (CC BY) license, which allows users to download, copy and build upon published articles, as long as the author and publisher are properly credited, which ensures maximum dissemination and a wider impact of our publications.

The book as a whole is distributed by MDPI under the terms and conditions of the Creative Commons license CC BY-NC-ND.

Contents

About the Editor vii

Cher-Ming Tan
Editorial for Special Issue on Reliability Analysis of Electrotechnical Devices
Reprinted from: *Appl. Sci.* 2022, 12, 4086, doi:10.3390/app12084086 1

Dipesh Kapoor, Cher Ming Tan and Vivek Sangwan
Evaluation of the Potential Electromagnetic Interference in Vertically Stacked 3D Integrated Circuits
Reprinted from: *Appl. Sci.* 2020, 10, 748, doi:10.3390/app10030748 7

Dipesh Kapoor, Vivek Sangwan, Cher Ming Tan, Vani Paliwal and Nirdosh Tanwar
Optimization of a T-Shaped MIMO Antenna for Reduction of EMI
Reprinted from: *Appl. Sci.* 2020, 10, 3117, doi:10.3390/app10093117 23

Yueh Chiang, Cher Ming Tan, Tsi-Chian Chao, Chung-Chi Lee and Chuan-Jong Tung
Investigate the Equivalence of Neutrons and Protons in Single Event Effects Testing: A Geant4 Study
Reprinted from: *Appl. Sci.* 2020, 10, 3234, doi:10.3390/app10093234 39

Yueh Chiang, Cher Ming Tan, Chuan-Jong Tung, Chung-Chi Lee and Tsi-Chian Chao
Lineal Energy of Proton in Silicon by a Microdosimetry Simulation
Reprinted from: *Appl. Sci.* 2021, 11, 1113, doi:10.3390/app11031113 53

Vimal Kant Pandey, Cherming Tan and Vivek Sangwan
GaN-Based Readout Circuit System for Reliable Prompt Gamma Imaging in Proton Therapy
Reprinted from: *Appl. Sci.* 2021, 11, 5606, doi:10.3390/app11125606 65

Yanruoyue Li, Xiaojun Yan, Guicui Fu, Bo Wan, Maogong Jiang and Weifang Zhang
Failure Analysis of SAC305 Ball Grid Array Solder Joint at Extremely Cryogenic Temperature
Reprinted from: *Appl. Sci.* 2020, 10, 1951, doi:10.3390/app10061951 79

Wendi Guo, Guicui Fu, Bo Wan and Ming Zhu
Reconstruction of Pressureless Sintered Micron Silver Joints and Simulation Analysis of Elasticity Degradation in Deep Space Environment
Reprinted from: *Appl. Sci.* 2020, 10, 6368, doi:10.3390/app10186368 95

Cher Ming Tan, Preetpal Singh and Che Chen
Accurate Real Time On-Line Estimation of State-of-Health and Remaining Useful Life of Li ion Batteries
Reprinted from: *Appl. Sci.* 2020, 10, 7836, doi:10.3390/app10217836 117

Luis Alberto Rodríguez-Picón, Luis Carlos Méndez-González, Roberto Romero-López, Iván JC Pérez-olguín, Iván Rodríguez-Borbón and Delia Julieta Valles-Rosales
Reliability Analysis Based on a Gamma-Gaussian Deconvolution Degradation Modeling with Measurement Error
Reprinted from: *Appl. Sci.* 2021, 11, 4133, doi:10.3390/app11094133 133

Cher-Ming Tan, Hsiao-Hi Chen, Jing-Ping Wu, Vivek Sangwan, Kun-Yen Tsai and Wen-Chun Huang
Root Cause Analysis of a Printed Circuit Board (PCB) Failure in a Public Transport Communication System
Reprinted from: *Appl. Sci.* 2022, 12, 640, doi:10.3390/app12020640 151

About the Editor

Cher Ming Tan

Cher Ming Tan, Professor received his Ph.D. in Electrical Engineering from the University of Toronto in 1992. He worked in the reliability field in the electronic industry before joining Nanyang Technological University as a faculty member (1996–2014). He joined Chang Gung University, Taiwan and set up a research Center on Reliability Sciences and Technologies in Taiwan and acts as Center Director. He is a Professor in the Electronic Department of Chang Gung University and Honorary Chair Professor at Ming Chi University of Technology, Taiwan. He has published 350+ international journal and conference papers and given many keynotes, talks and tutorials at international conferences. He has written more than seven books and chapters in the field of reliability. He is an Editor of five international reputed journals. He is on the committee of IEEE EDS Reliability Physics. He is a past chair of IEEE Singapore Section, senior member of IEEE and ASQ, and Distinguished Lecturer of IEEE Electronic Device Society on reliability. He was the first individual recipient of the Ishikawa-Kano Quality Award in Singapore in 2014. He is also listed in the top 2% of scientists in the world by Stanford University.

Editorial

Editorial for Special Issue on Reliability Analysis of Electrotechnical Devices

Cher-Ming Tan

Center for Reliability Science and Technology, Chang Gung University, Wenhua 1st Road, Guishan District, Taoyuan City 33302, Taiwan; cmtan@cgu.edu.tw

1. Introduction

Advancement in electrotechnical devices has indeed revolutionize our daily lives. It will be hard to imagine going through our daily lives without cell phones, computers, and internets, to name a few. Consequently, our reliance on electrotechnical devices has been increasing significantly, and thus the dependability of these devices is becoming crucial.

Dependability of a device is determined by its reliability and its constituting components therein. Reliability is defined as the probability that a product, system, or service will perform its intended function adequately for a specified period of time or will operate in a defined environment without failure. The component of reliability is as depicted in Figure 1 to the following: The components of reliability include probability, durability, dependability, availability and quality over time [1].

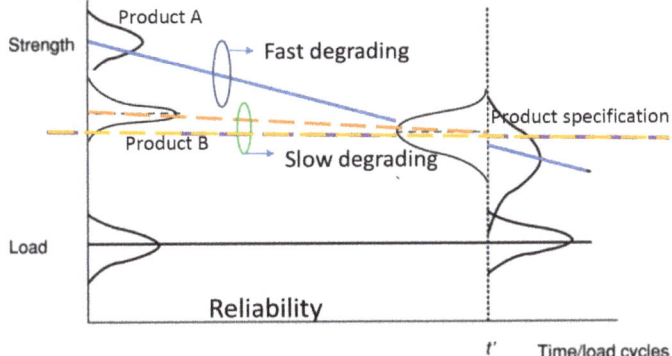

Figure 1. Different between Reliability and Current Quality Practice.

Although reliability is part of quality as can be learnt from the definition of quality according to ISO8402, which stated that quality is the totality of features and characteristics of a product or service that bear on its ability to satisfy stated or implied needs [2]. These stated or implied needs of course include the operation of products at prolonged period or its mission lifetime. However, due to the impossibility to check the reliability of every single product, as reliability tests usually involve time-consuming and destructive measurements, quality assurance can only assure the quality practice of the incoming materials, manufacturing processes through SPC, Cpk and yield, and product outcoming quality check against specifications. This renders a risk in the reliability of the products as depicted in Figure 1.

Figure 1 shows the strength distributions of two product batches, A and B. As the strength of the products in batch A is well above the specification, its quality at time zero is consider excellent whilst products in batch B suffers some yield loss. In both cases, their strengths are above the expected load to be applied to them, and hence they can operate without failure. During continuous operations, the strengths of products will degrade as expected. Thus, their mean strengths will be reduced, but the spread of the strength distribution increases since not all the products in a batch will degrade equally. If the degradation rate of products in batch A is much higher than that of batch B, some products in batch A will go below the load at time t', and some failure will occur. Such failures are considered as reliability failure. On the other hand, there is no failure for products in batch B, although its yield is not 100%. This figure explains the limitations of the current quality assurance methodology as degradation rate can only be found through reliability or accelerated life test. Even if reliability test is performed but with no failure, degradation rate will not be able to determine either.

However, the determination of the degradation rate can be time consuming and cost intensive as the number of failures in a selected accelerated life test must be sufficient to provide a good estimation of the degradation rate of a batch. Furthermore, the choice of accelerated life test must correspond to the dominant failure mechanisms of the products in normal field operating conditions, and this requires detail failure analysis of field failure if the product is already in the market or a good understanding of the possible various failure mechanisms before the products are launched.

To address the above-mentioned, and in view of the short time to market available for products, a new research area has recently emerged where the physics of failures and material sciences are employed to explore the various possible degradation mechanisms with the aid of powerful simulation tools. Statistical concept is also incorporated in the simulation to show the possible variation in the time to failure and reliability statistics is employed for the extrapolation of time to failure from accelerated test conditions to different field operating conditions. Such research area is called reliability sciences where experimentation, multi-scale simulation, atomistic finite element modelling and statistics are used concurrently, so that reliability of components can be predicted before the assembly of the final product. In this way, the reliability of a product will be determined beforehand, allowing for faster and cheaper modifications, before 'final' products are developed. This is a paradigm shift in product reliability assurance, and current reliability evaluation, when performed on the final product, would then act as secondary measurements, confirming the initially predicted reliability [3].

This special issue focusses on the exploration of the underlying degradation sciences. We focus on technical important devices, including antenna for IoT, 3D stack-chips for today high-performance integrated circuit assembly, Lithium-Ion battery (LiB) for electric vehicle and energy storage, devices in radiation environment such as space applications as the demand for satellites increases for IoT applications and medical applications, and printed circuit board assembly for wireless communication in public transport. As accelerated life tests for some highly reliable products can be too long, degradation analysis is employed. Improving the accuracy of the degradation test data analysis with respect to the measurement errors is also included in this special issue.

2. Electromagnetic Interference for Stack-Chips and Antenna

3D stack-chips are becoming common due to the increasing complexity of circuits and the requirement of small outline for an integrated circuit. Various factors can affect their reliability such as the stress induced from TSV [4], thermal dissipation effectiveness in the dense and 3D structure of the chips [5]. One of the advantages of using 3D stack-chip is the low latency of the interconnections for high-speed operation [6]. However, such high-speed operation can also induce electromagnetic field that can couple to the dices above and below, due to their proximity, resulting in serious electromagnetic interference

(EMI) between the dices. Dipesh et al. first examined such issue using electromagnetic computation method apply to integrated circuit.

They found that near field electromagnetic measurement is ineffective to identify the hot spots of electromagnetic emission within the stacked dies, and they thus developed a computational method to identify the electromagnetic emission hot spots in integrated circuits with experimental verification. They found that if the stack distance between dices is too small, the coupled magnetic field strength can be as high as 16 times as compared with an unstacked IC. Thus, a minimum distance between the stack must be maintained. The method they developed can be carried out at the IC design stage before fabrication, enabling optimization of the stack distance as well as circuit design for reducing the EMI, and eliminating the significant cost of fabrication owing to improper design.

Using the method developed, they also identified the key dimensional parameters of a miniaturized single element of T-shaped MIMO antenna which is commonly used today for data communication. They found an optimized antenna with 1 GHz improvement in the bandwidth due to the improved return loss that provides new opportunities for the antenna's utilization in different applications. This optimized antenna has lower current distribution that gives lower power dissipation, and its 1 m sphere EMI was also reduced.

These are examples of using computational power to ensure good performances and lower power dissipation, which in turn can lead to better durability.

3. Electronic Devices under Radiation Environment

Radiation on electronics is no longer limit to space application only. Advances in semiconductor technology have enhanced the functionality of integrated circuits (ICs) with reduced feature sizes down to nm. Proton and neutron radiation are always present around us but have not been detrimental to electronics at sea level. With the decreasing feature size of transistors and the increasing density of transistors in ICs, the effect of radiation on the reliability of semiconductor devices, sensors, and their electronic circuits (collectively called electronic systems) is no longer negligible, even at sea level. ISO26262 International safety standard states the requirements for automotive electronics to be radiation immune to a certain level. Thus, radiation reliability of electronics is an emerging area that cannot be neglected [7].

Although the dominant radiation source at sea level for electronics is neutron, neutron radiation test is rarely available due to the control difficult for neutron beam. Chiang et al. investigated the possibility of having proton to replace neutron for electronics using Monte Carlo simulation, and they found that 200 MeV protons closely resemble the effect from neutron radiation, and this concur with the suggestion from NASA, USA. They also used the Monte Carlo simulation to demonstrate the invalidity of the conventional concept of using linear energy transfer (LET) of radiation particles to estimate the single event effect (SEE) in electronic devices. Instead, lineal calculation of energy lost per step for each specific track should be used for the prediction of microelectronic devices failure due to SEEs. With such Monte Carlo simulation, they managed to predict the radiation reliability of microelectronics devices.

The increasing application of radiation therapy in medical treatment and diagnosis has also subjected the medical electronics to radiation. X-ray is a familiar example which subjects electronics to photon radiation, and the emerging proton therapy is another example which subjects electronics to stray proton radiation. Silicon devices are vulnerable to radiation, and to improve the reliability of the radiation system with good electronic control, Gallium Nitride (GaN) devices are suggested. Vimal et al. developed a GaN-based readout system in proton therapy treatment system. In particular, they designed an operation amplifier (OPA) which is a basic circuit module for the electronic in the readout system using GaN and simulation results indicated good performances. The proposed OPA was also configured for different applications such as transimpedance amplifier, integrator, and the adder which are needed in the prompt gamma readout system. Simulation results show successful operation for these applications. When these different applications are put together, a complete GaN-based prompt gamma readout circuit is implemented, and the result shows successful processing of the prompt gamma signal where its energy and position of the proton beam in a human body can be accurately provided for subsequent digital conversion and information extraction.

The increasing demand of IoT also prompt the significant increase in the need of satellites for communication. Radiation effects should be carefully monitored for the reliability of the electronics of satellites in the outer space, but radiation hardened electronics are too costly for IoT applications. In addition to the electronics components used, the solder joints of components onto printed circuit board also need to be reliable. It is found that the reliabilities of the commonly used Sn/Pb and SnAgCu solder joints are not good under the thermal shock conditions in the outer space. Guo et al. proposed an environmentally friendly interconnection material, namely pressureless sintered micron silver which is economical. Before predicting the reliability of this material in the proposed applications, the degradation mechanism of this material under the deep space environment was studied. They employed finite element analysis (FEA) to study the degradation physics, and their results were verified with their experiments. Their studies showed an elasticity degradation during the thermal shock, which induces lattice shift and dislocation slip in the microstructure. The dislocation piling up causes the effective elastic modulus of sintered joints to decline due to broken atomic bonds. With the understanding of the degradation physics, degradation model can be developed to extrapolate the lifetime of such material from accelerated test condition to predict its reliability and suitability in deep space usage.

The increasing important of electronics operating under radiation environment can be seen from the above description, and from the above few works, one can see that there are still a lot more work needed in this area.

4. Remaining Useful Life (RUL) of Lithium-Ion Battery (LiB)

While product lifetime can be predicted through reliability tests in most cases, field reliability is likely different. A typical rule of thumb is to expect the unreliability in the field to be twice what was observed in the lab. This is because product will usually receive harsher treatment in the field than in the lab. Units being tested in the labs are often carefully set up and adjusted by engineers prior to the beginning of the test. The tests are performed by trained technicians who are adept at operating the product being tested. Most end-use customers do not have the training and experience in its operation, thus leading to many more operator-induced failures than would be experienced during in-house testing [8].

In some cases, accelerated life test cannot be performed such as LiB and other highly reliable products. This is because accelerated test condition for LiB can affect the chemistry in LiB and the degradation mechanisms will be completely different from that operate at normal condition. For highly reliable products, the test time is simply too high and accelerated degradation tests usually employed.

Regardless of the situations, ability to estimate the remaining useful life of a product, as it might have been through various operating conditions, will be useful. Tan et al. developed a methodology to estimate the RUL for LiB. Their method combines the electrochemistry-based electrical model and semi-empirical capacity fading model on a discharge curve of a LiB for the estimation of its maximum stored charge capacity, and thus its state of health (SoH). The method developed produces a close form that relates SoH with the number of charge-discharge cycles as well as operating temperatures and currents, and its inverse application allows us to estimate the remaining useful life of LiB for a given SoH threshold level. The estimation time is less than 5 s as the combined model is a closed-form model, and hence it is suitable for real time and on-line applications.

5. Degradation Analysis

With the increasing product reliability, prediction of lifetime through accelerated life test can be too long, and degradation analysis is usually employed. However, in most degradation tests, the measuring processes may cause variation in the observed measures.

As the measuring process is inherent to the degradation testing, it is important to establish schemes that define a certain level of permissible measurement error such that a robust reliability estimation can be obtained. Rodriguez-Picon et al. proposed an approach to deal with this measurement error based on a deconvolution operation. An illustrative example based on a fatigue-crack growth dataset is presented to illustrate the applicability of the proposed scheme.

6. Failure Analysis

Failure analysis is common in electronic industry; however, the depth of analysis needs to be further to uncover the underlying science of the failure mode. Physics of failure is needed to ensure product reliability as only then root cause can be identified. Tan et al. demonstrated a step-by-step failure analysis methodology for multilayer printed circuit boards that led an observed failure mode to the root cause with verification of the root cause and the effectiveness of the corresponding corrective action. This printed circuit board (PCB) is used in public transport systems. With the identified root cause, a modified processing method was developed and the observed failure mode was no longer observed with this modified method, which verified the root cause. The PCBs with the modified processing method also underwent appropriate reliability tests and the corrective action of this modified processing method is confirmed.

Li et al. performed detail failure analysis on failure obtained from low temperature and shock test that mimic deep space exploration. They found that the failure was in the Pb-free (Sn-3.0Ag-0.5Cu) solder joint, and the low temperature changes the fracture characteristic of Sn-3.0Ag-0.5Cu (SAC305) from ductileness to brittleness. The crack occurred at solder joints from the stress loaded by shock test. When the crack reached a specific length, the failure occurred. A verification test was conducted to verify the failure mechanism. The transition temperature range of SAC305 was also confirmed as -70--$-80\ °C$.

7. Future Outlook

Reliability of products is an implicit expectation from users, but the current quality assurance method is limited to provide sufficient information to assure product reliability. Reliability tests on products are also facing challenges as mentioned earlier, and a paradigm shift is on the reliability science. This is an important and emerging area in reliability, and it requires professional from different disciplines to come together and share their works so that we can gel the different aspects of reliability science together. With the increasing variety of electrotechnical devices in increasing application areas, many research works are expected. We look forward to seeing the explosion in this area.

Funding: This research received no external funding.

Acknowledgments: This issue would not be possible without the contributions of various authors, hardworking and professional reviewers, and dedicated editorial team of applied Sciences. Congratulations to all authors. I would like to take this opportunity to show my sincere gratefulness to all reviewers. Finally, I like to express my gratitude to the editorial team of Applied Sciences that make this work possible.

Conflicts of Interest: The author declares no conflict of interest.

References

1. What Is Reliability? Available online: https://asq.org/quality-resources/reliability (accessed on 7 April 2022).
2. *ISO 8402:1994*; Quality—Vocabulary. Available online: https://www.saiglobal.com/pdftemp/previews/osh/as/as8000/8400/8402.pdf (accessed on 7 April 2022).
3. Physics of Degradation: The Difference between Reliability Engineering and Reliability Science. Available online: https://researchoutreach.org/articles/physics-degradation-difference-between-reliability-engineering-reliability-science/ (accessed on 7 April 2022).
4. Jiang, T.; Im, J.; Huang, R.; Ho, P.S. Through-silicon via stress characteristics and reliability impact on 3D integrated circuits. *Mrs Bull.* **2015**, *40*, 248–256. [CrossRef]

5. Salvi, S.S.; Jain, A. A Review of Recent Research on Heat Transfer in Three-Dimensional Integrated Circuits (3-D ICs). *IEEE Trans. Compon. Packag. Manuf. Technol.* **2021**, *11*, 802–821. [CrossRef]
6. Philip, G.; Christopher, B.; Peter, R. (Eds.) *Handbook of 3D Integration*; Wiley-VCH Verlag GmbH & Co., kGaA: Weinheim, Germany, 2011.
7. Tan, C.M.; Pandey, V.K.; Chiang, Y.; Lee, T.P. Electronic Reliability Analysis under Radiation Environment. *Sens. Mater.* **2022**, *34*, 1119–1131. [CrossRef]
8. Reliability Hot Wire, Issue 6. August 2001. Available online: https://weibull.com/hotwire/issue6/hottopics6.htm (accessed on 7 April 2022).

Article

Evaluation of the Potential Electromagnetic Interference in Vertically Stacked 3D Integrated Circuits

Dipesh Kapoor [1], Cher Ming Tan [1,2,3,4,*] and Vivek Sangwan [5]

1. Center for Reliability Sciences & Technologies and Electronic Engineering Department, Chang Gung University, Taoyuan 33302, Taiwan; dipeshkapoor.dk@gmail.com
2. Institute of Radiation Research, College of Medicine of Chang Gung University, Taoyuan 33302, Taiwan
3. Department of Mechanical Engineering, Ming Chi University of Technology, New Taipei City 24301, Taiwan
4. Department of Urology, Chang Gung Memorial Hospital, Linkou, Taoyuan 33302, Taiwan
5. Center for Reliability Sciences & Technologies, Chang Gung University, Taoyuan 33302, Taiwan; sangwanvivek81@gmail.com
* Correspondence: cmtan@cgu.edu.tw

Received: 4 December 2019; Accepted: 16 January 2020; Published: 21 January 2020

Featured Application: Three-dimensional integrated circuit (3D-IC) is the trend for future C development because of its small form factor, high performance, low cost, and heterogeneous integration in system-in-package technologies. Several issues associated with 3D-IC besides their fabrication processes are being addressed, such as heat dissipation and so on, however, the electromagnetic interference between stacks has not been considered. In view of the high operating frequency, high power requirements of today IC, and together with the decreasing separation between the stacked dies, investigation on the subject matter is necessary, as they will soon become important. Unfortunately, the study of electromagnetic interference (EMI) within 3D-IC is difficult from the measurement, as elaborated in this paper. In this work, we develop a simulation methodology and show the importance of EMI evaluation for 3D-IC, and we also illustrate the presence of minimum separation between the stacked dies for the avoidance of EMI.

Abstract: Advancements in the functionalities and operating frequencies of integrated circuits (IC) have led to the necessity of measuring their electromagnetic Interference (EMI). Three-dimensional integrated circuit (3D-IC) represents the current advancements for multi-functionalities, high speed, high performance, and low-power IC technology. While the thermal challenges of 3D-IC have been studied extensively, the influence of EMI among the stacked dies has not been investigated. With the decreasing spacing between the stacked dies, this EMI can become more severe. This work demonstrates the potential of EMI within a 3D-IC numerically, and determines the minimum distance between stack dies to reduce the impact of EMI from one another before they are fabricated. The limitations of using near field measurement for the EMI study in stacked dies 3D-IC are also illustrated.

Keywords: 3D-IC (three-dimensional integrated circuit); electromagnetic interference; near field measurement

1. Introduction

Three-dimensional stack-dies integrated circuit (3D-IC) is a technology used for multi-functionality and high-speed circuit devices [1]. 3D-IC has received much interest recently because of its small form factor, high performance, and heterogeneous integration in system-in-package technologies [2–4]. Three-dimensional technologies can provide higher chip-to-chip bandwidths at lower power levels that cannot be accomplished with available conventional packaging techniques. Additionally, they also offer the potential for low cost through heterogeneous integration. Interposers offer the first step to 3D integration of ICs, which simplifies physical planning and thermal management [4]. To achieve higher performance and reliability in 3D-IC, new design rules have to be developed because of the specific electrical, mechanical, and thermal constraints of 3D stacks [5–7].

Although stacking of chips began from memory and later on extend to logic chip, the technology of stacking is now also extending to radio frequency applications, which consist of mixed-signal IC to benefit from the 3D integration including shorter propagation delay; full isolation between analog and digital circuits in mixed-signal 3D-IC [8,9]; and lower parasitic, which can reduce power consumption. With the 5G chip, this will be a trend.

Despite the above-mentioned advantages, there are several challenges associated with 3D-ICs. The two most significant challenges are the quality of the wafer–wafer bonding and thermal management. Three-dimensional technology poses a major challenge in thermal management owing to the increase in power density and number of vertically stacked active layers [10,11]. Therefore, researchers have made efforts to solve for the thermal stress issue in 3D-ICs through thermally aware floor planning or task scheduling [12].

Besides these challenges, there is another emerging challenge that has been rather overlooked, namely the electromagnetic coupling between stack dies [13–15]. These electromagnetic couplings are known as electromagnetic interferences (EMIs), which are a consequence of high frequency switching currents, complex power delivery paths, and an increase in parasitic couplings in comparison with 2D integration.

The increase in the parasitic couplings between stacked dies is the result of the decreasing stack up distances between them, as shown in Figure 1, as extracted from the works of [16–23]. In 2017, this distance decreased to 15 µm [16] from 150 µm in year 2000 [22], which is a 10-fold decrement. This decrease is necessary in order to maximize the advantages of 3D vertical integration including shorter interconnect lengths, greater integration density, and lower power consumption, which enhances the overall performance of the system with such technologies. The main driving forces to keep the stack distance lesser between the stacked dies are the increased demand of dies with faster data exchange rate, lower power consumption, and smaller size, and such a decreasing trend is expected to continue in the future.

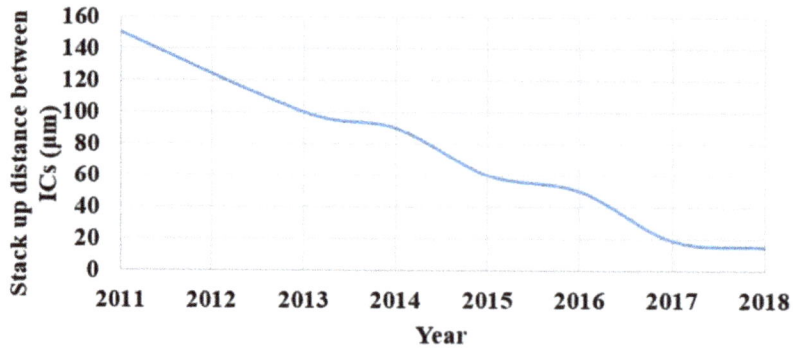

Figure 1. Three-dimensional stack-dies integrated circuits (3D-ICs) trend for vertical distance between ICs [16–23].

With this decreasing trend of the distance between the dies, one limitation will be the thermal dissipation of the 3D-IC, and another one will be the electromagnetic interference between them. If one of the dies in the 3D-IC stack is the source of EMI, it could affect the nearby dies when the distance between them is small. This work explores the latter limitation.

This limitation depends on two factors. One is the circuit susceptible to EMI and another one is the parasitic coupling. The former one requires ingenious circuit design, which is beyond the scope of this work.

To reduce the parasitic coupling with the decreasing distance between the dies, several possibilities are available. Adding of ground plane between them or adjusting the placement of vias to conduct away the electromagnetic field are some examples. Irrespective of the methods employed, a method to evaluate the EMI in the stack before fabrication of the 3D stack dies will be necessary.

Near-field (NF) measurement is the basic for the detection of EMI at IC level. In the work of [24], the correlation between near field scanning and electrostatic discharge susceptibility measurement of the CPU IC is studied, which is an initial study for the application of NF measurement. In another paper [25], a procedure is shown to characterize radiated EMI and nearby conducting objects using near-field scanning measurements. Yu et al. [26] proposed a model from near-field measurement to calculate the magnitude and phase of the dipoles. In the work of [27], Slattery showed that near field scanning can be utilized to calculate far field from the IC. ICs are often the primary source of radiated emissions, and near field magnetic field can help engineers to track down EMI culprit and solve the problems with a robust IC design [28].

With the well-known rapid decrease of the intensity of EM wave with the distance [29], if the source for high EMI is from the die in the middle of the stack of a 3D-IC, such near field measurement will not be able to pinpoint the source of high EMI. Therefore, a computational method is required, and if the computation can be performed before the fabrication of the dies, the cost and time for re-design can be saved significantly, and it is this necessity that forms the motivation of this work.

While the equation of the intensity of electromagnetic field strength versus distance is well known, the 3D distribution of the field strength in the presence of various materials within each die, as well as the 2D distribution of electromagnetic field over a given circuit on a die, are to be determined in order to evaluate the EMI. This work proposes a simulation method from a given GDSII file for such evaluation, and we aim to provide the evaluation before the fabrication of the 3D-IC.

The example studied in this work on the stacking of power amplifier ICs is expected to be the future, and research is being done on the thermal management [30–33]. While thermal management research is being considered, the impact of electromagnetic interference (EMI) among the stack dies has not been studied. This work represents the first of its kind. Advancements in the functionalities and operating frequencies of integrated circuits (IC) have led to the necessity of measuring their EMI. The stacking of power amplifier chips in this work can be considered as an extreme case so as to bring out the important of the EMI consideration in stack dies more vividly, and it is also a good example of generalization of ICs with multi functionality of communication, high performance, and high frequency IC technology. The methodology developed here can be applied easily to other cases of stack dies.

2. Integrated Circuit under Study

This work employs the gallium nitride high electron mobility transistor (GaN-HEMT) power amplifier IC as an example, which operates on frequency range of 2–4 GHz (S-band of electromagnetic spectrum). This IC possess the current advancements in frequency and power, which makes it a potential candidate to study the effect of EMI in this work. To illustrate the methodology developed here, we use a hypothetical 3D-IC where we stack up two and three of this same IC to form the 3D-IC in this work, and the severity of EMI in the stacked dies is also computed.

An optical image of the power amplifier IC under study at 1600× magnification is shown in Figure 2. Its input power (Pin) is 28 dBm and the corresponding output power (Pout) is 30.3 dBm. Drain to source voltage (VDS) of the transistors is 26 V with the gate to source voltage (VGS) of −2.8 V. The dimensions of the IC are 1700 μm × 1400 μm × 107.92 μm.

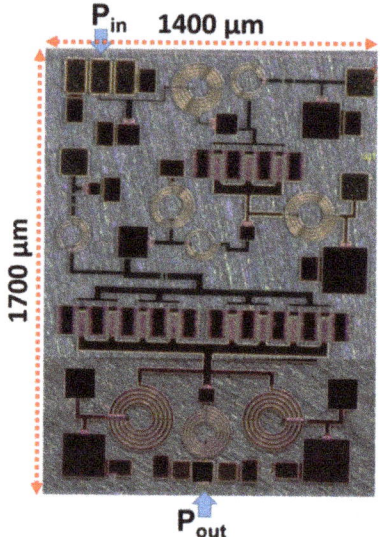

Figure 2. Image of power amplifier IC at 1600× magnification captured (using Leica S8 APO equipment).

3. Simulation Methodology

Recently, Tan et al. [34] developed a method to "fabricate" an IC from its GDSII file in a computer under ANSYS environment. Once the IC is fabricated, the ANSYS high frequency structure simulator (HFSS) is employed to compute the EM field distribution over the surface of the chip. ANSYS HFSS has a limitation that it cannot simulate transistors. To overcome the limitation of ANSYS HFSS, transistors are removed from the layout and respective voltages and currents as obtained from circuit simulator are inputted at respective circuit nodes for the proper functionality of the IC. With such replacements, our circuit can be completed, and the currents thus computed will be accurate. In this work, the values of these voltages and currents are obtained from the advanced design systems (ADSs) file of the circuit during the circuit simulation in the post layout phase. Figure 3 shows the flow chart of the method as also reported in the work of [28] for clarity, and Figure 4 shows the "fabricated" IC in the ANSYS environment. With this method, the magnetic field distribution of the IC can be obtained at different frequencies (2, 3, and 4 GHz). The method is verified with experimental results, as shown in the work of [34].

In Figure 4a, black circles represent input ports, while Pin is another input port used in simulation in HFSS.

The following equations are used in determining the electromagnetic distribution as given below [35]:

$$\nabla \times (\frac{1}{\mu_r}\nabla \times E(x,y,z)) - k_0^2 \varepsilon_r E(x,y,z) = 0, \quad (1)$$

$$H = \frac{\nabla \times E}{-j\omega\mu}, \quad (2)$$

$$\nabla \times (\frac{1}{\mu_r}\nabla \times E(x,y,z)) - (k_0^2 \varepsilon_r - jk_0 Z_0 \sigma)E(x,y,z) = 0, \quad (3)$$

$$E(x,y,z) = \int_S (j\omega\mu_0 H_{\tan}G + E_{\tan} \times \nabla G + E_{normal}\nabla G)dS, \quad (4)$$

where

$E(x,y,z)$ is a phasor representing an oscillating electric field;
k_0 is the free space wave number;
$\omega \sqrt{\mu_0 \varepsilon_0}$, where ω is the angular frequency, which is $2\pi f$;
μ_r is the complex relative permeability;
ε_r is the complex relative permittivity;
S represents the radiation boundary surfaces;
j is the imaginary unit ($\sqrt{-1}$);
ε_0 is the relative permeability of the free space;
H_{tan} is the component of the magnetic field that is tangential to the surface;
E_{normal} is the component of the electric field that is normal to the surface;
E_{tan} is the component of the electric field that is tangential to the surface;
G is the free space Green's function;
J is the current density.

Equations (1)–(3) are employed in the finite element method (FEM). FEM solves for the three-dimensional electromagnetic field, taking x, y, and z directions in consideration, and solves for the volume distribution using the curl function. As we are dealing with 3D IC, FEM in ANSYS HFSS will be most appropriate for the determination of EM distribution.

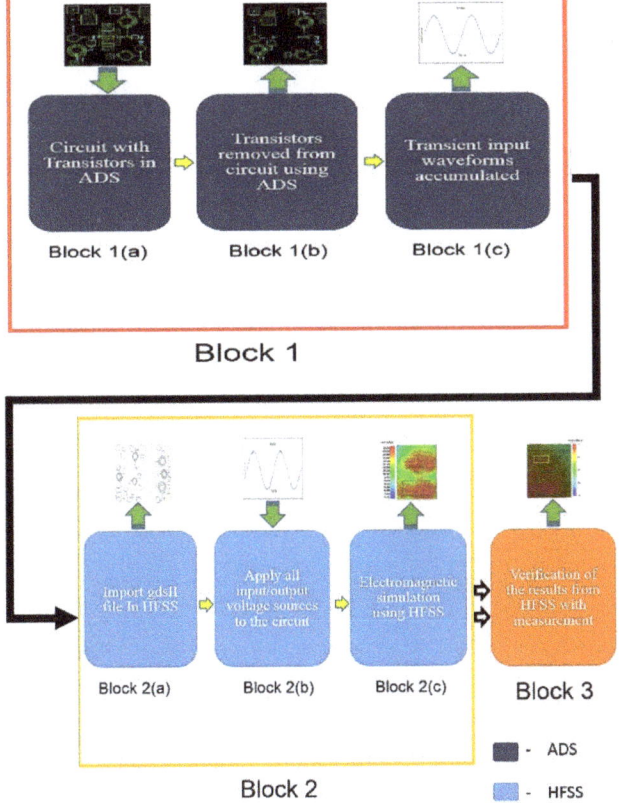

Figure 3. Flow chart of the methodology followed for simulation [28]. ADS, advanced design system; HFSS, high frequency structure simulator.

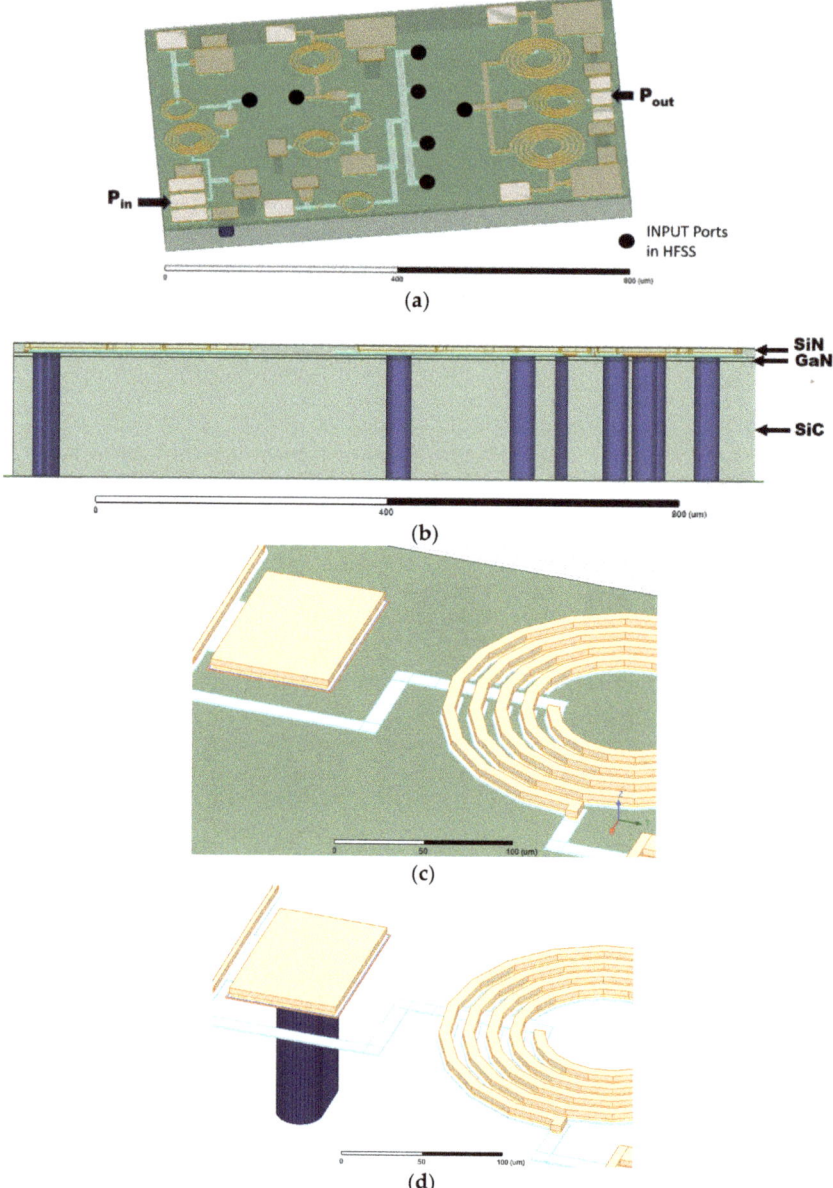

Figure 4. Power amplifier IC layout in HFSS: (**a**) 3D isometric view where black circles represent input ports, (**b**) side-view with dielectric layers information about material, (**c**) zoomed in view of layout with substrate, and (**d**) zoomed in view of layout without substrate (blue color represents via). GaN, gallium nitride; SiC, silicon carbide; SiN, silicon nitride.

To perform electromagnetic field calculation, FEM is necessary. This necessity can be seen by comparing the magnetic field simulation using HFSS (which utilizes FEM) and SIwave (which uses method of moments (MOM)). Figure 5a shows the test structure used for magnetic field simulation and Figure 5b shows the results. One can see a significant difference in the amplitude and distribution of the magnetic field between the two methods. As solved Maxwell equations in FEM (3D method) will be more accurate, MOM (2.5D method) should be used with caution.

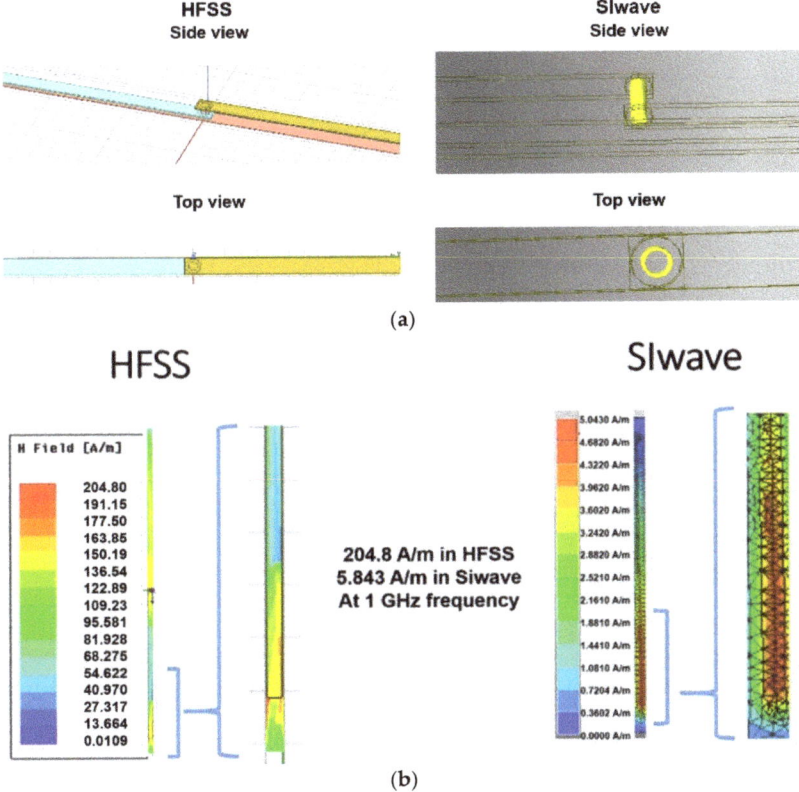

Figure 5. Comparison of method of moments (MOM) and finite element method (FEM) simulation: (**a**) side and top view of test structure used for simulation, (**b**) magnetic field distribution in HFSS (3D method) and SIwave (2.5D method) software, respectively.

The impact of the H-field is more significant than electric field emissions in vicinity of the device, as wave impedance for the H-field is small and increases with the distance from the source [36]. After some distance from the source, wave impedances of the H-field and electric field become same, which is found to be 318 mm for the present case, as calculated using the Equation in the work of [36]. However, near field measurements in the present scenario are conducted at 0.2 mm, and thus magnetic field analysis becomes important. For the effect of EMI on 3D-IC circuits, the distance is even smaller, and hence magnetic field computation is the focus of this work.

Dielectric layers used in ANSYS HFSS simulation methodology are silicon carbide (SiC), GaN, and silicon nitride (SiN) to resemble the actual IC. Perfect electric conductor (PEC) boundary condition is used for ground plane and lumped port excitation is provided [34]. The overall electromagnetic simulation takes around 20 min.

Figure 6 shows the simulation result for 3 GHz operating frequency, respectively. Such a simulation method has been verified experimentally in our other works [34].

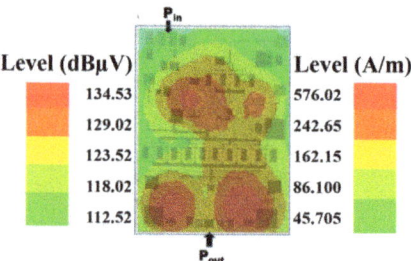

Figure 6. Magnetic field distribution obtained through HFSS simulation over the gallium nitride high electron mobility transistor (GaN-HEMT) power amplifier IC at 3 GHz operating frequency.

4. Electromagnetic Field Intensity Distribution

Figure 7 shows the magnetic field strength of the device under test plotted at various vertical distances and operating frequencies from the surface of the IC. The maximum magnetic field strength of 1.07×10^5 A/m, at 3 GHz operating frequency, is present at the top surface of the IC, and it decreases to 4.60×10^4 A/m drastically within a very small change in distance (around 1 μm) from the surface of the IC. At 40 μm vertically from the surface of the IC, the maximum magnetic field strength is reduced to approximately 10^3 A/m, and it is reduced to around 10^2 A/m at 140 μm above the surface. This is expected as the magnetic field is inversely proportional to square of the distance according to inverse square law [27]. Therefore, with the current commercial near field measurement at 200 μm from the surface, the severity of EMI from an IC can be underestimated, but this can be unimportant as the EMI strength decreases rapidly outside the package. However, such EMI can affect the nearby circuits within a chip and has yet been overlooked. Either these nearby circuits can be on the same die, or they can be on the stacked die, if they are close. If the strong EME source is within the stack dies instead of the topmost die, its identification will not be possible with the current near field measurement. In fact, even if the hot spot for EME is on the topmost die, its identification will be difficult, as illustrated below.

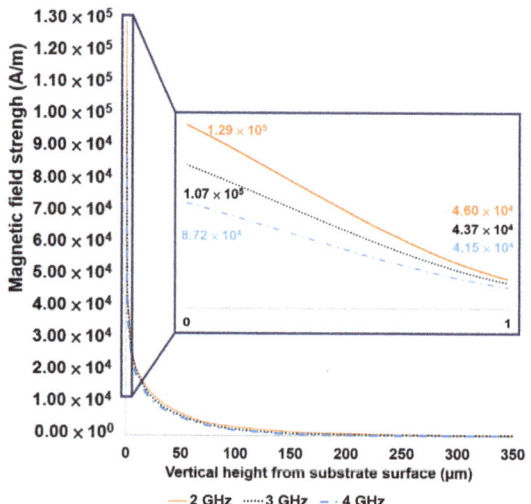

Figure 7. Maximum magnetic field strength versus vertical height of reference plane at different operating frequencies (2, 3, and 4 GHz).

Figure 8a shows the magnetic field distribution at the surface of the IC, where the spot for high EME can be identified with high precision. As the observational plane moves away from the surface, the area of the hot spot disperse, and this dispersion renders it difficult to identify the actual source of EME precisely. At 50 µm, several hot spots of EME appear as a result of the dispersion, and this can result in tackling the wrong parts of a circuit to reduce the EMI. Such a situation will be very likely for stacked dies in 3D. In other words, even if high EME exists on a die, localization of hot spot through electromagnetic field measurement will not be possible.

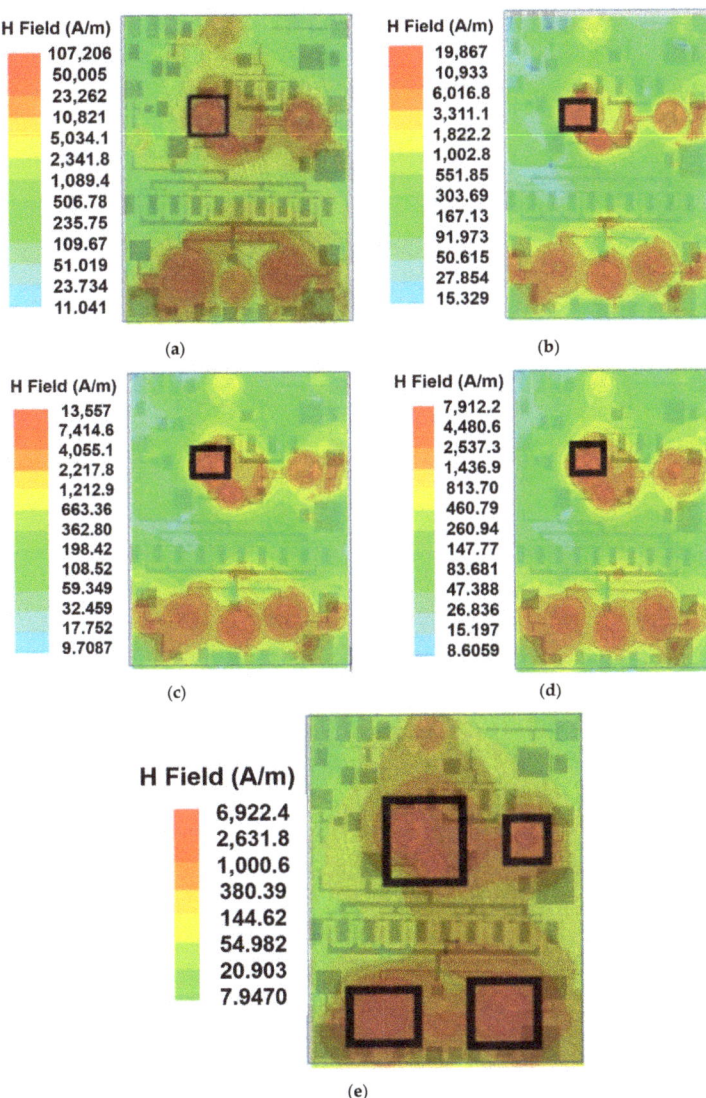

Figure 8. Maximum magnetic field strength for an operating frequency of 3 GHz at different vertical heights: (**a**) 0 µm (black box represents the maximum emission region), (**b**) 10 µm, (**c**) 20 µm, (**d**) 40 µm, and (**e**) 50 µm.

In fact, the above results indicate that the vertical distance for near field (NF) scanning should be in the region from 10 to 40 µm. However, below 10 µm, the equipment required to perform NF measurement is not available and, beyond 40 µm, the precision to find the source of EMI is low. Thus, simulation of EM field strength is necessary. In practical measurement, the probe distance is usually set at 200 µm to prevent the damage to the probe during scanning.

As the electromagnetic field strength decreases significantly over the distance, one can determine the minimum distance between dies separation for 3D-ICs, so that the effect of EMI from a die will have minimum effect on other dies in a stack. To investigate this minimum separation between the dies, simulation is performed on a hypothetical 3D-IC using GaN-HEMT power amplifier IC, as shown in Figure 9, with two dies (2D-3D-IC) and three dies (3D-3D-IC) aligned vertically to each other. The vertical separation between the dies (denoted as 'X') varies from 25 µm to 300 µm in this work.

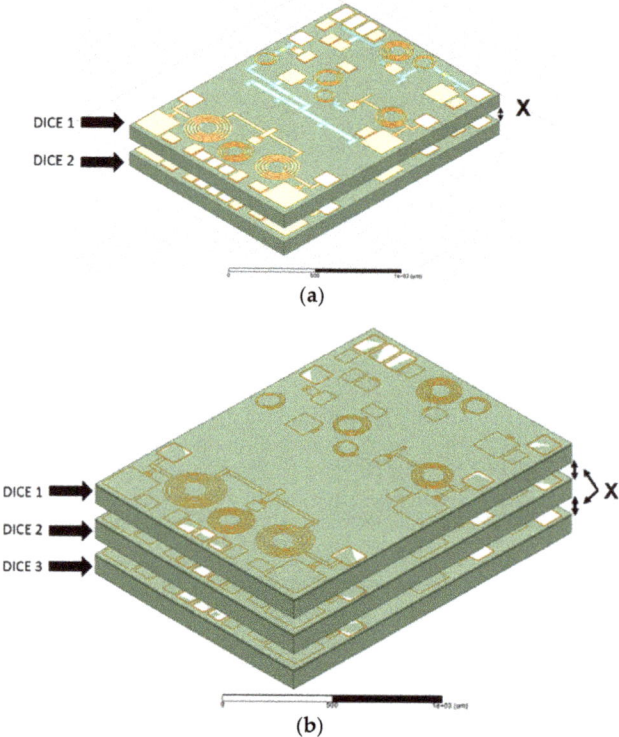

Figure 9. Hypothetical 3D-IC layout in HFSS: (**a**) two layers in stack of 3D-IC (2D-3D-IC) and (**b**) three layers in stack of 3D-IC (3D-3D-IC).

Figure 10 shows the magnetic field distributions on the surface of the topmost die as the distance between the two dies vary in 2D-3D-IC. Change in 'X' changes the magnitude of the magnetic field distribution significantly, as expected, especially when the distance between them is very small. The distribution pattern remains the same, as all the operational conditions are identical.

Figure 11 shows that the maximum magnetic field strength for 2D-3D-IC at the surface is 7.44×10^5 A/m, where the red line represents the maximum magnetic field strength for single IC at the surface, which is 1.07×10^5 A/m. We are seeing more than double in the maximum magnetic field strength with an additional die. This value decreases rapidly as we increase the dies' separation, but it is always more than 1.07×10^5 A/m, showing the significant effect of EMI from one die to another in a stack. The EMI increases up to seven times when two dies stack up.

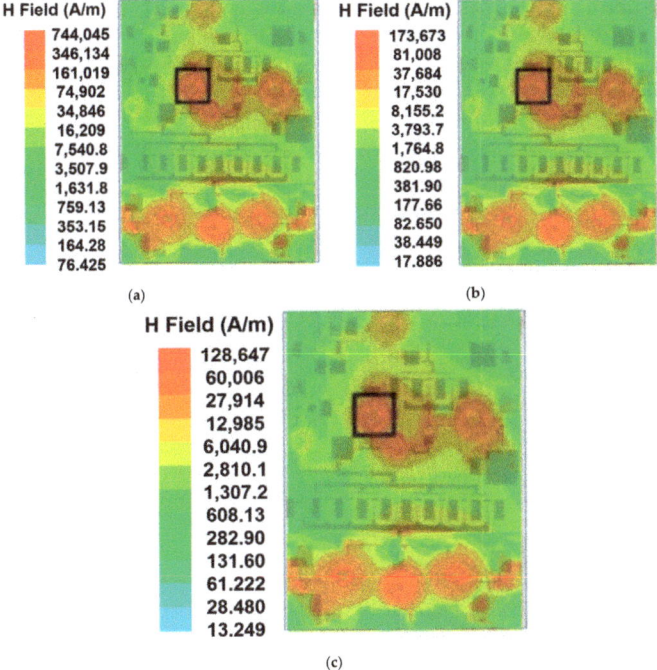

Figure 10. Maximum magnetic field strength at the surface of top IC with different vertical separation between the dies in a stack of 3D-IC with two layers. The vertical separation is as follows: (**a**) 0 µm, (**b**) 50 µm, and (**c**) 300 µm, respectively, and the operating frequency is 3 GHz.

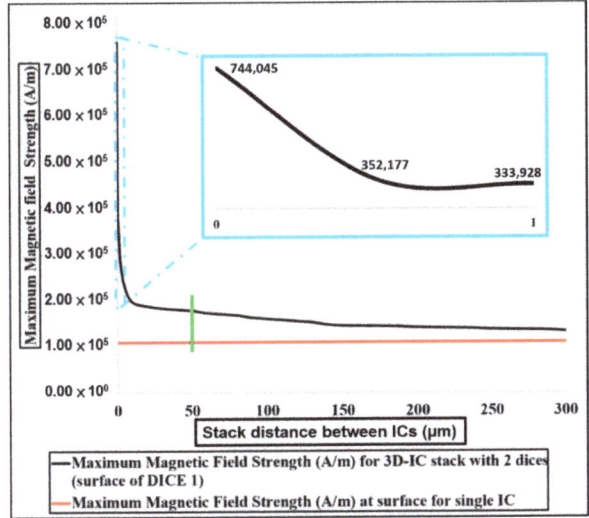

Figure 11. Maximum magnetic field strength versus the distance between the layers of 2D-3D-IC at 3 GHz operating frequency.

Figure 12 shows the difference in the maximum magnetic field for 2D-3D-IC and 3D-3D-IC, where the black line represents the maximum magnetic field strength for single IC at the surface, which is 1.07×10^5 A/m. The maximum magnetic field strength for 3D-3DIC structure obtained is 4.28×10^6 A/m. Although this maximum value decreases as we increase the dies' separation, it will always be greater than 7.44×10^5 A/m, which is maximum magnetic field strength of the 2D-3D-IC structure. In other words, the more dies are stacked up, the higher will be the maximum magnetic field strength. Hence, it could be possible that each die in a stack may not have high EME, and the stacked structure could have very high EME that renders circuit malfunction.

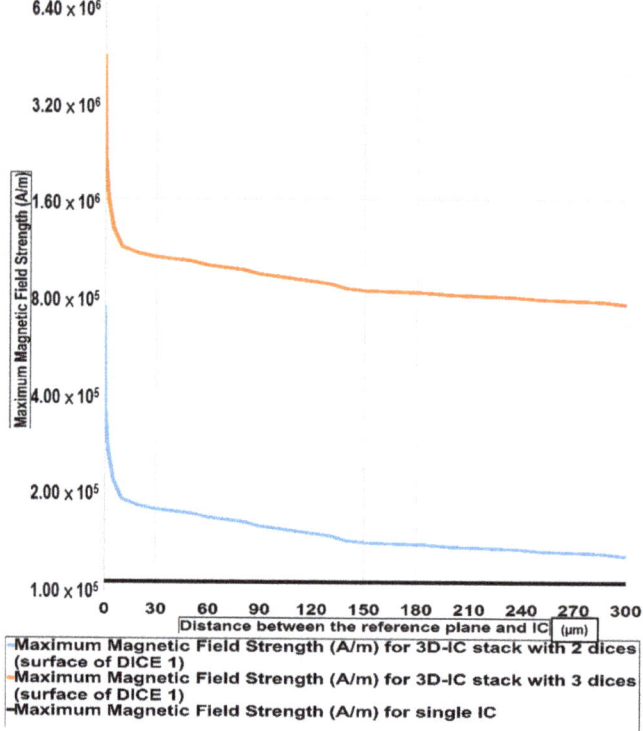

Figure 12. Maximum magnetic field strength versus the distance between the reference plane and topmost layer of 3D-IC at 3 GHz operating frequency.

Figure 13 shows the maximum magnetic field strength on the top surface of each die in the 3D-3D-IC stack, and one can see that the maximum field strength occurs at the top surface of the middle die. This is expected as it is placed in the middle of the two dies and it gets the emissions from the top and bottom dies.

With the decrease of 'X' in 3D-ICs, an increase in the EMI can be expected. Tradeoff is thus required to keep the emissions and distance in check, and it requires careful evaluation, which is demonstrated in this work.

For the example used in this work, the optimum distance between stack of dies should be around 50 µm for 2D-3D-IC, while this value increases to 105 µm for 3D-3D-IC, shown by green lines in Figures 11 and 13, respectively. These values are considered because further increase in 'X' does not result in a significant reduction in the magnetic field. If the lowest EMI is needed, then 'X' should be even more, but there is a limit on the minimum magnetic field achievable, which is higher with a higher number of stacked dies.

The lowest EMI in 3D-IC structure is always higher than the single IC, as can be seen in Figures 12 and 13, and thus reduction of EMI from a 3D-IC architecture is critical.

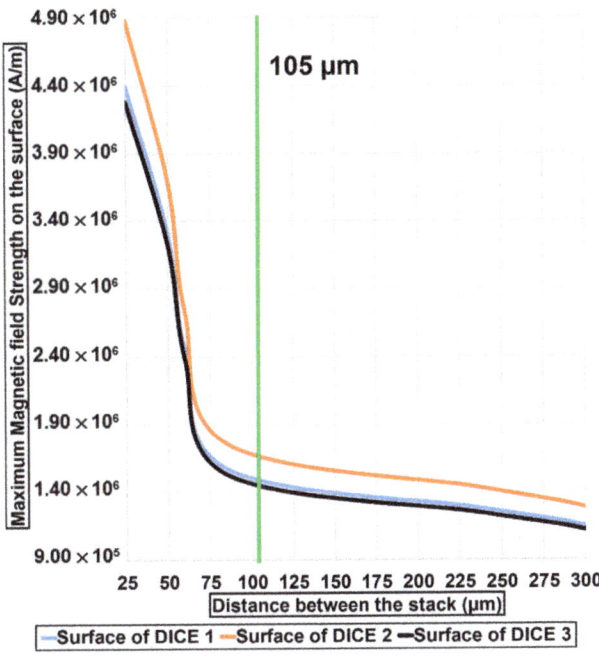

Figure 13. Maximum magnetic field strength versus the distance between the layers of 3D-IC at 3 GHz operating frequency at the top surface of all the dies.

5. Conclusions

Integrated circuits for high-speed applications are prone to electromagnetic interference (EMI) issues. For detecting these EMIs, near field electromagnetic measurement is an important tool, as it is helpful in determining the locations of high electromagnetic emission (EME hot spots) within the IC. As the 3D stack dies technology is increasingly important owing to its several advantages, the EMI issue within the stacked dies must be considered when the stack dies are high frequency and high power.

However, near field electromagnetic measurement is ineffective to identify the hot spots of electromagnetic emission within the stacked dies, as illustrated in this work, and a computational method is developed to evaluate and identify the EME hot spots in integrated circuits with experimental verification. Using this method, we found that the EMI issue within the stack dies is a lot more severe. This work shows that the minimum value of the maximum magnetic field strength for 3D-stacked IC, as the distance between the stack increases, is about six times higher than an un-stacked IC. If the stack distance is smaller, the magnetic field strength can be as high as 16 times as compared with an unstacked IC. Thus, a minimum distance between the stack must be maintained. The actual minimum distance depends on the electromagnetic susceptible of the designed circuit. The method developed in this work can be done at the IC design stage before fabrication, enabling optimization of the stack distance as well as circuit design for reducing the EMI, and eliminating the significant cost of fabrication owing to improper design.

Author Contributions: Conceptualization, C.M.T.; Methodology, D.K.; Software, D.K.; Validation, D.K.; Formal analysis, V.S.; Investigation, D.K.; Resources, C.M.T.; Data curation, V.S.; Writing—original draft preparation, D.K.;

Writing—review and editing, C.M.T.; Visualization, D.K.; Supervision, C.M.T.; Project administration, C.M.T.; Funding acquisition, C.M.T. All authors have read and agreed to the published version of the manuscript.

Funding: This work was funded by the Chang Gung University research Grant QZRPD123 and CIRPD2F0024.

Conflicts of Interest: The authors declare no conflict of interest.

References

1. Zhan, C.J.; Juang, J.Y.; Lin, Y.M.; Huang, Y.W.; Kao, K.S.; Yang, T.F.; Lu, S.T.; Lau, J.H.; Chen, T.H.; Lo, R.; et al. Development of fluxless chip-on-wafer bonding process for 3DIC chip stacking with 30 μm pitch lead-free solder micro bumps and reliability characterization. In Proceedings of the 2011 IEEE 61st Electronic Components and Technology Conference (ECTC), Lake Buena Vista, FL, USA, 31 May–3 June 2011; pp. 14–21. [CrossRef]
2. Knickerbocker, J.U.; Andry, P.S.; Dang, B.; Horton, R.R.; Interrante, M.J.; Patel, C.S.; Polastre, R.J.; Sakuma, K.; Sirdeshmukh, R.; Sprogis, E.J.; et al. Three—Dimensional silicon integration. *IBM J. Res. Dev.* **2008**, *52*, 553–569. [CrossRef]
3. Dang, B.; Wright, S.L.; Andry, P.S.; Sprogis, E.J.; Tsang, C.K.; Interrante, M.J.; Webb, B.C.; Polastre, R.J.; Horton, R.R.; Patel, C.S.; et al. 3D chip stacking with C4 technology. *IBM J. Res. Dev.* **2008**, *52*, 599–609. [CrossRef]
4. Emma, P.G.; Kursun, E. Is 3D chip technology the next growth engine for performance improvement? *IBM J. Res. Dev.* **2008**, *52*, 541–552. [CrossRef]
5. Chaabouni, H.; Rousseau, M.; Leduc, P.; Farcy, A.; Thuaire, R.E.F.A.; Haury, G.; Valentian, A.; Billiot, G.; Assous, M.; de Crecy, F.; et al. Investigation on TSV impact on 65nm CMOS devices and circuits. In Proceedings of the 2010 International Electron Devices Meeting, San Francisco, CA, USA, 6–8 December 2010; pp. 796–799. [CrossRef]
6. Cherman, V.O.; de Messemaeker, J.; Croes, K.; Dimcic, B.; van der Plas, G.; de Wolf, I.; Beyer, G.; Swinnen, B.; Beyne, E. Impact of through silicon vias on front-end-of-line performance after thermal cycling and thermal storage. In Proceedings of the 2012 IEEE International Reliability Physics Symposium (IRPS), Anaheim, CA, USA, 15–19 April 2012; pp. 5–9. [CrossRef]
7. Beyne, E. Electrical, thermal and mechanical impact of 3D TSV and 3D stacking technology on advanced CMOS Devices-Technology directions. In Proceedings of the 2011 IEEE International 3D Systems Integration Conference (3DIC), Osaka, Japan, 31 January–2 February 2012. [CrossRef]
8. Yahalom, G.; Ho, S.; Wang, A.; Ko, U.; Chandrakasan, A. Analog-Digital Partitioning and Coupling in 3D-IC for RF Applications. In Proceedings of the 2016 IEEE International 3D Systems Integration Conference (3DIC), San Francisco, CA, USA, 8–11 November 2016; pp. 1–4.
9. Uemura, S.; Hiraoka, Y.; Kai, T.; Dosho, S. Isolation Techniques Against Substrate Noise Coupling Utilizing Through Silicon Via (TSV) Process for RF/Mixed Signal SoCs. *IEEE J. Solid-State Circuits* **2012**, *47*, 810–816. [CrossRef]
10. Li, F.; Nicopoulos, C.; Richardson, T.; Xie, Y.; Narayanan, V.; Kandemir, M. Design and management of 3D chip multiprocessors using network-in-memory. *Proc. Int. Symp. Comput. Archit.* **2006**, *2006*, 130–141. [CrossRef]
11. Loi, G.L.; Agrawal, B.; Srivastava, N.; Lin, S.C.L.S.C.; Sherwood, T.; Banerjee, K. A thermally-aware performance analysis of vertically integrated (3-D) processor-memory hierarchy. In Proceedings of the 43rd annual Design Automation Conference, San Francisco, CA, USA, 24–28 July 2016; pp. 991–996. [CrossRef]
12. Mizunuma, H.; Yang, C.L.; Lu, Y.C. Thermal modeling for 3D-ICs with integrated microchannel cooling. In Proceedings of the 2009 International Conference on Computer-Aided Design, San Jose, CA, USA, 2–5 November 2009; pp. 256–263. [CrossRef]
13. Koo, K.; Lee, S.; Kim, J. Vertical noise coupling on wideband low noise amplifier from on-chip switching-mode DC-DC converter in 3D-IC. In Proceedings of the IEEE Electromagnetic Compatibility of Integrated Circuits (EMC Compo), Dubrovnik, Croatia, 6–9 November 2011; pp. 35–40.
14. Sicard, E.; Jianfei, W.; Shen, R.; Li, E.P.; Liu, E.X.; Kim, J.; Cho, J.; Swaminathan, M. Recent Advances in Electromagnetic Compatibility of 3D-ICs-Part II. *IEEE Electromagn. Mag.* **2016**, *5*, 65–74. [CrossRef]
15. Sicard, E.; Jianfei, W.; Shen, R.J.; Li, E.P.; Liu, E.X.; Kim, J.; Cho, J.; Swaminathan, M. Recent advances in Electromagnetic Compatibility of 3D-ICs-Part I. *IEEE Electromagn. Mag.* **2015**, *4*, 79–89. [CrossRef]

16. Agrawal, A.; Huang, S.; Gao, G.; Wang, L.; DeLaCruz, J.; Mirkarimi, L. Thermal and Electrical Performance of Direct Bond Interconnect Technology for 2.5D and 3D Integrated Circuits. In Proceedings of the 2017 IEEE 67th Electronic Components and Technology Conference (ECTC), Orlando, FL, USA, 30 May–2 June 2017; pp. 989–998. [CrossRef]
17. Chen, S.; Tzeng, P.; Hsin, Y.; Wang, C.; Chang, P.; Chen, J.; Chen, T.; Hsu, T.; Chang, H.; Zhan, C.; et al. Implementation of Memory Stacking on Logic Controller by Using 3DIC 300mm Backside TSV Process Integration. In Proceedings of the 2016 International Symposium on VLSI Technology, Systems and Application (VLSI-TSA), Hsinchu, Taiwan, 25–27 April 2016; Volume 2, pp. 1–2. [CrossRef]
18. Zhang, C.; Thadesar, P.; Zia, M.; Sarvey, T.; Bakir, M.S. Au-NiW mechanically flexible interconnects (MFIs) and TSV integration for 3D interconnects. In Proceedings of the 2014 International 3D Systems Integration Conference (3DIC), Kinsdale, Ireland, 1–3 December 2014; pp. 1–4. [CrossRef]
19. Chen, J.C.; Chen, E.H.; Tzeng, P.J.; Lin, C.H.; Wang, C.C.; Chen, S.C.; Hsu, T.C.; Chen, C.C.; Hsin, Y.C.; Chang, P.C.; et al. Low-cost 3DIC process technologies for Wide-I/O memory cube. In Proceedings of the 2015 International Symposium on VLSI Technology, Systems and Applications, Hsinchu, Taiwan, 2–3 June 2015. [CrossRef]
20. Yang, C.C.; Hsieh, T.Y.; Huang, W.H.; Wang, H.H.; Shen, C.H.; Shieh, J.M. Sequentially stacked 3DIC technology using green nanosecond laser crystallization and laser spike annealing technologies. In Proceedings of the 2015 IEEE 22nd International Symposium on the Physical and Failure Analysis of Integrated Circuits, Hsinchu, Taiwan, 29 June–2 July 2015; pp. 389–391. [CrossRef]
21. Madhour, Y.; Zervas, M.; Schlottig, G.; Brunschwiler, T.; Leblebici, Y.; Thome, J.R.; Michel, B. Integration of intra chip stack fluidic cooling using thin-layer solder bonding. In Proceedings of the 2013 IEEE International 3D Systems Integration Conference (3DIC), San Francisco, CA, USA, 2–4 October 2013; pp. 1–8. [CrossRef]
22. Ki, W.M.; Kang, M.S.; Yoo, S.; Lee, C.W. Fabrication and bonding process of fine pitch Cu pillar bump on thin Si chip for 3D stacking IC. In Proceedings of the 2011 IEEE International 3D Systems Integration Conference (3DIC), Osaka, Japan, 31 January–2 February 2012; pp. 3–6. [CrossRef]
23. Kang, S.; Cho, S.; Yun, K.; Ji, S.; Bae, K.; Lee, W.; Kim, E.; Kim, J.; Cho, J.; Mun, H.; et al. TSV optimization for BEOL interconnection in logic process. In Proceedings of the 2011 IEEE International 3D Systems Integration Conference (3DIC), Saka, Japan, 31 January–2 February 2012. [CrossRef]
24. Ahmad, H.; Izadi, O.H.; Shinde, S.; Pommerenke, D.; Shumiya, H.; Maeshima, J.; Araki, K. A study on correlation between near-field EMI scan and ESD susceptibility of ICs. In Proceedings of the IEEE International Symposium on Electromagnetic Compatibility Signal/Power Integrity, Washington, DC, USA, 7–11 August 2017; pp. 169–174. [CrossRef]
25. Jin, S.; Cracraft, M.A.; Zhang, J.; DuBroff, R.E.; Slattery, K. Using near-field scanning to predict radiated fields. International Symposium on Electromagnetic Compatibility. In Proceedings of the 2004 International Symposium on Electromagnetic Compatibility (IEEE Cat. No.04CH37559), Silicon Valley, CA, USA, 9–13 August 2004; Volume 1, pp. 14–18. [CrossRef]
26. Yu, Z.; Koo, J.; Mix, J.A.; Slattery, K.; Fan, J. Extracting physical IC models using near-field scanning. In Proceedings of the IEEE International Symposium on Electromagnetic Compatibility, Fort Lauderdale, FL, USA, 25–30 July 2010; pp. 317–320. [CrossRef]
27. Slattery, K.P. A comparison of the near field and far field emissions of a Pentium (R) clock IC. In Proceedings of the IEEE EMC International Symposium. Symposium Record. International Symposium on Electromagnetic Compatibility, Montreal, QC, Canada, 13–17 August 2001; Volume 1, pp. 547–551. [CrossRef]
28. Xiaopeng, D.; Deng, S.; Hubing, T.; Beetner, D. Analysis of chip-level EMI using near-field magnetic scanning. In Proceedings of the International Symposium on Electromagnetic Compatibility, Silicon Valley, CA, USA, 9–13 August 2004; Volume 1, pp. 174–177. [CrossRef]
29. Stefanini, I.; Markovic, M.; Perriard, Y. 3D Inductance and Impedance Determination Taking Skin Effect into Account. In Proceedings of the IEEE International Conference on Electric Machines and Drives, San Antonio, TX, USA, 15 May 2005; Volume 2, pp. 74–79. [CrossRef]
30. Lau, J.H.; Yue, T.G. Thermal management of 3D IC integration with TSV (through silicon via). In Proceedings of the 2009 59th Electronic Components and Technology Conference, San Diego, CA, USA, 26–29 May 2009; pp. 635–640.

31. Patrick, L.; de Crecy, F.; Fayolle, M.; Charlet, B.; Enot, T.; Zussy, M.; Jones, B.; Barbe, J.C.; Kernevez, N.; Sillon, N.; et al. Challenges for 3D IC integration: Bonding quality and thermal management. In Proceedings of the 2007 IEEE International Interconnect Technology Conferencee, Burlingame, CA, USA, 4–6 June 2007; pp. 210–212.
32. Kiran, P.; Loh, G.H. Thermal analysis of a 3D die-stacked high-performance microprocessor. In Proceedings of the 16th ACM Great Lakes Symposium on VLSI, Providence, RI, USA, 16–18 May 2010; pp. 19–24.
33. Lau John, H.; Yue, T.G. Effects of TSVs (through-silicon vias) on thermal performances of 3D IC integration system-in-package (SiP). *Microelectron. Reliab.* **2012**, *52*, 2660–2669.
34. Sangwan, V.; Kapoor, D.; Tan, C.M.; Lin, C.H.; Chiu, H.C. High Frequency Electromagnetic Simulation and Optimization for GaN-HEMT Power Amplifier IC. *IEEE Trans. Electromagn. Compat.* **2018**, *61*, 564–571. [CrossRef]
35. Ansys Inc. *Ansys HFSS Technical Notes*; ANSYS: Canonsburg, PA, USA, 2017.
36. Ott, H.W. *Electromagnetic Compatibility Engineering*; John Wiley Sons: Hoboken, NJ, USA, 2009.

© 2020 by the authors. Licensee MDPI, Basel, Switzerland. This article is an open access article distributed under the terms and conditions of the Creative Commons Attribution (CC BY) license (http://creativecommons.org/licenses/by/4.0/).

Article

Optimization of a T-Shaped MIMO Antenna for Reduction of EMI

Dipesh Kapoor [1], Vivek Sangwan [2], Cher Ming Tan [1,3,4,5,*], Vani Paliwal [6] and Nirdosh Tanwar [7]

1. Center for Reliability Sciences & Technologies and Electronic Engineering Department, Chang Gung University, Taoyuan 33302, Taiwan; dipeshkapoor.dk@gmail.com
2. Center for Reliability Sciences & Technologies, Chang Gung University, Taoyuan 33302, Taiwan; sangwanvivek81@gmail.com
3. Institute of Radiation Research, College of Medicine of Chang Gung University, Taoyuan 33302, Taiwan
4. Center of Reliability Engineering, Ming Chi University of Technology, New Taipei City 24301, Taiwan
5. Department of Urology, Chang Gung Memorial Hospital, Linkou, Taoyuan 33302, Taiwan
6. Electronics and Communication Department, DIT University, Dehradun, Uttarakhand 248009, India; vanipaliwal19@gmail.com
7. Department of Engineering, Lexorbis Consulting Private Ltd., New Delhi 110001, India; nirdoshtanwar@gmail.com
* Correspondence: cmtan@cgu.edu.tw

Received: 7 February 2020; Accepted: 27 April 2020; Published: 29 April 2020

Featured Application: A four-element wideband multiple-input multiple-output (MIMO) antenna is optimized using a genetic algorithm (GA). The antenna consists of a stub, four reduced ground planes, and four T-shaped radiating elements. The behavior of the antenna is analyzed with the following parameters: return loss, electromagnetic interference (EMI), and operating frequency. The antenna design is applicable to different applications. A 2^K factorial design is used to identify the key design parameters responsible for affecting the above-mentioned performance indexes of the antenna. A GA optimization technique is utilized to increase the frequency range, where the return loss is less than −10 dB, while its EMI is reduced to within a 1 m sphere of radiation, so as to reduce the threat of the antenna to nearby devices. The increased frequency range also makes it suitable for various applications such as satellite communication, imaging, and radar communication.

Abstract: In this paper, optimization of a miniaturized multiple-inputs multiple-outputs (MIMO) antenna was performed. This antenna was composed of a T-shape radiating element with stub and reduced ground plane and a compact size of 25 mm × 25 mm × 1.6 mm. The behavior of antenna was evaluated in terms of return loss (S-parameter < −10 dB), electromagnetic interference (EMI), and operation frequency. The antenna design is applicable to many applications. A 2^K factorial design combined with a genetic algorithm (GA) optimization technique were used to identify the key design parameters responsible for affecting the performance quality of the antenna. Optimization of the antenna design for EMI reduction was utilized, and the optimal design showed enhanced bandwidth of the antenna and reduced power consumption.

Keywords: factorial design of experiment; genetic algorithm optimization; return loss; electromagnetic interference; multiple-input multiple-output (MIMO)

1. Introduction

Antennas are used in several aspects of life, from human body networks to space communication. There are many antennas around us, and hence their interference with one another and with other electronic gadgets must be reduced. Electromagnetic interference (EMI) between transmitting and receiving antennas is an important consideration for relative antenna placement [1]. Whether onboard or offboard, radiating and receiving electronic appliances such as antennas can interfere with other such devices and can reduce their application effectiveness.

Multipath propagation issues arise in conventional transmission systems due to signal degradation in the transmission medium of the transmitter and receiver. To overcome this issue, multiple-input multiple-output (MIMO) technology can be implemented [2–8]. A variety of research on MIMO antennas can be found in the literature with an aim of achieving high quality, isolated signals along with compact antenna size to overcome the aforementioned issues.

Tian et al. [9] and Liu et al. [10] discussed channel capacity to enhance antenna efficiency by reducing the mutual coupling and effects of spatial correlation under the Rayleigh fading channel assumption. Abdul et al. [11] and Khalighi et al. [12] proposed a method to identify the number of antennas at the asymmetric base station as well as in mobile units in order to increase the effectiveness of the station. Du et al. [13] addressed the need to optimize overall MIMO system capacity, which includes the unequal costs of antennas at both channels.

In antenna design optimization, algorithms inspired by natural processes have been applied, and they are genetic algorithms (GAs) [14–17], particle swarm optimization (PSO) [18,19], and their variants [20,21]. Genetic algorithm (GA)-based optimization algorithms have been utilized for antenna array positioning and have been successful in finding optimized antenna schemes [22]. These optimization algorithms do not consider the assumptions about the design, and the generation of optimal parameters is necessary to meet the defined design criteria [23].

In this work, we wanted to optimize the MIMO antenna design with respect to its EMI. The factors that affect the EMI can be classified into its structural parameters and the material properties of the substrate. This work focused on the structural parameters only. However, there are ten structural parameters, as we discuss later, making optimization challenging.

To improve the effectiveness of optimization, 2^K factorial design methodology was applied in this work. This methodology is commonly employed for many experimental designs and has been shown to be the most effective experimental design methodology [24]. The 2^K refers to designs with K factors where each factor has just two levels. Through this methodology, we can screen a large number of factors and identify only those parameters that are important so that the scale of optimization can be reduced.

In this work, a four-element wideband MIMO antenna was optimized. This antenna consisted of a stub, reduced ground plane, and T-shaped radiating elements, and its size was 25 mm × 25 mm × 1.6 mm. Optimization was done using 2^K factorial and GA optimization algorithms, with Minitab and ANSYS High Frequency Structure Simulator (HFSS) software, respectively. EMI was considered the parameter to be optimized, and this optimization also improved the return loss and bandwidth, as will be shown in this work. Optimization helps in reducing the interference and enhancing the operating frequency and bandwidth of the antenna [25], which makes the proposed antenna suitable for various applications such as radar communication, imaging, and satellite communication.

2. Antenna Design and Fabrication

The design diagram of a single-element MIMO antenna is shown in Figure 1. This MIMO antenna is designed on RT/Duroid substrate with relative permittivity of 2.2 and dielectric loss tangent (tanδ) of 0.0009 [26], having a cross-section of 25 mm × 25 mm × 1.6 mm. A microstrip feed line of 50 ohms is utilized to provide the input supply to the antenna.

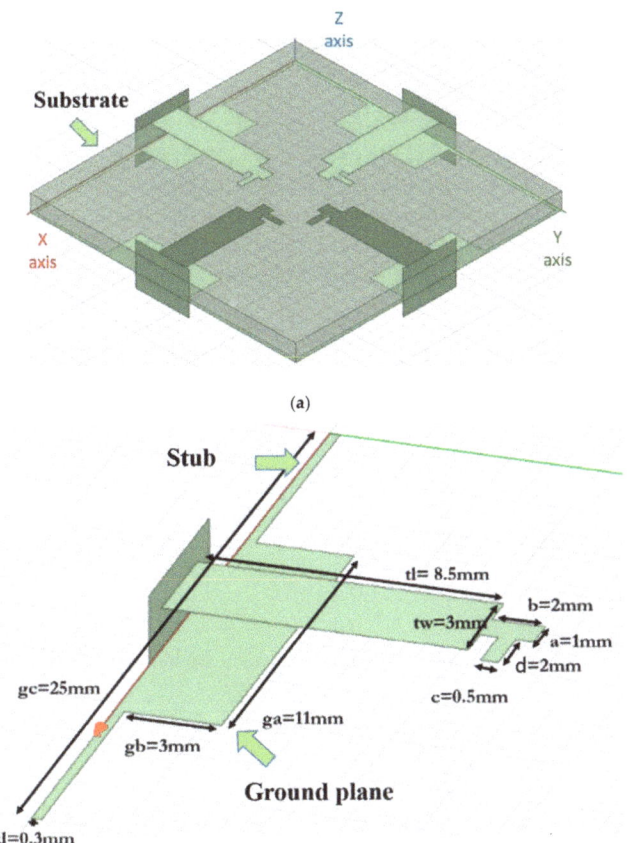
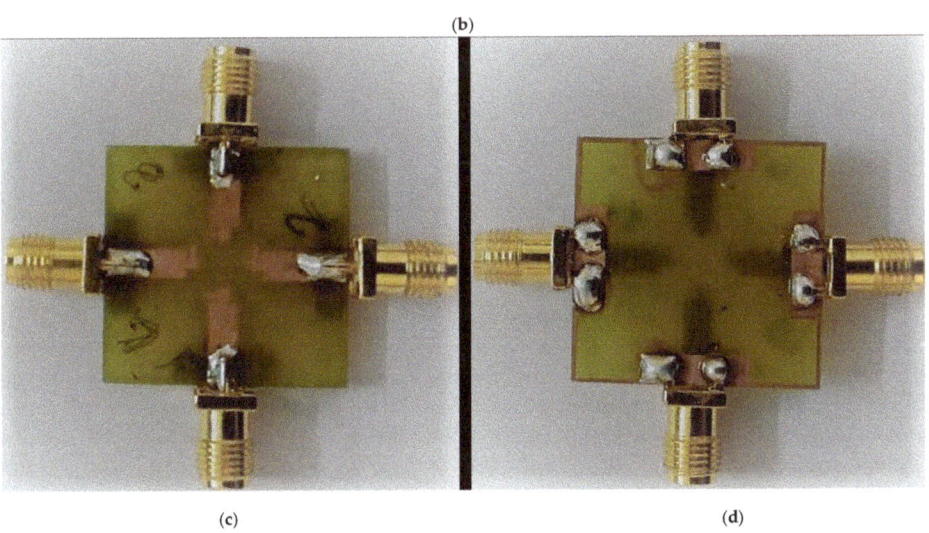

Figure 1. (a) Multiple-input multiple-output MIMO antenna with substrate, (b) dimensional geometry of one element in the four-element MIMO antenna, (c) top view of the fabricated antenna, and (d) back view of the ground plane. Substrate is absent in (b) to highlight the ground plane.

The reduced ground plane was designed at the bottom of the substrate and straight stubs of 0.3 mm width are used to establish a connection with the reduced ground plane. Figure 1a shows the MIMO antenna with the substrate understudy in ANSYS HFSS V19.0, and Figure 1b shows the structural parameters of the T-shaped antenna. Initial values of design parameters were a = 1 mm, b = 2 mm, c = 0.5 mm, d = 2 mm, tw = 3 mm, tl = 8.5 mm, ga = 11 mm, gb = 3 mm, gc = 25 mm, and gd = 0.3 mm. Figure 1c,d shows the top and back views of the fabricated MIMO antenna.

Figure 2 shows the simulated and measured return loss of the initial antenna design as shown in Figure 1. The difference between simulated and measured results may be attributed to the fabrication discrepancies and associated tolerances because of the variation of dielectric constant (ε_r) and loss tangent (tanδ) of the RT/Duroid substrate with frequency and relative humidity [27–29].

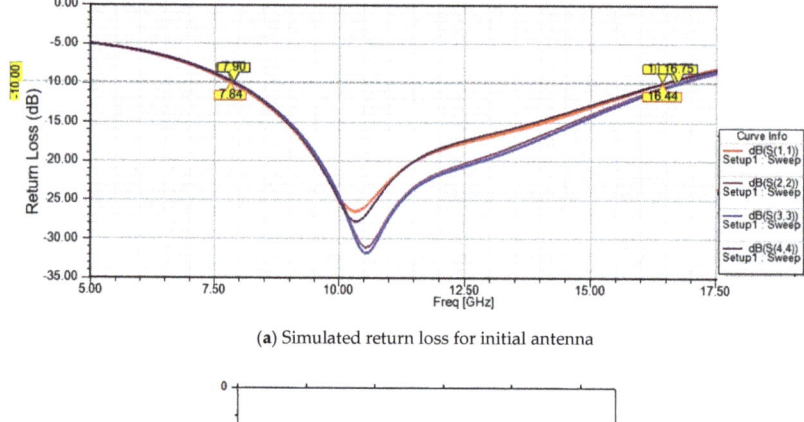

(**a**) Simulated return loss for initial antenna

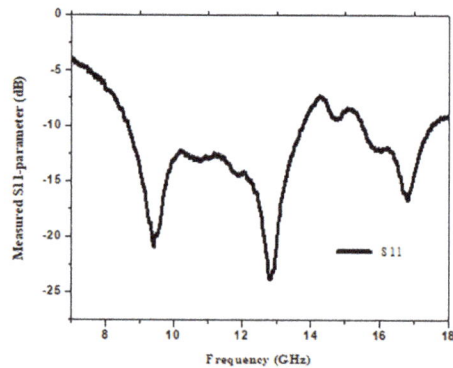

(**b**) Measured return loss for initial antenna

Figure 2. Return loss of initial antenna (**a**) simulated and (**b**) measured.

A small change in design, relative permittivity, and thickness of substrate were incorporated into our antenna simulation in order to investigate the causes of the difference between the simulated and measured results. Figure 3 shows the modifications made to the antenna design, where the area under the black eclipse represents an area of change, while the red box represents a zoomed-in area. The changes were made in the angle of the patch in the antenna design (to mimic the actual antenna), dielectric constant value to 2.22 from 2.2, and substrate thickness to 1.575 mm from 1.6 mm. All these minor changes resulted in the matching of the losses between the simulated and measured results. However, negligible changes were observed in EMI, radiation pattern, peak gain, and surface current distribution, as shown in Figure 4. Hence, we proceeded with our ideal antenna design instead of the physical antenna design that contained non-ideality.

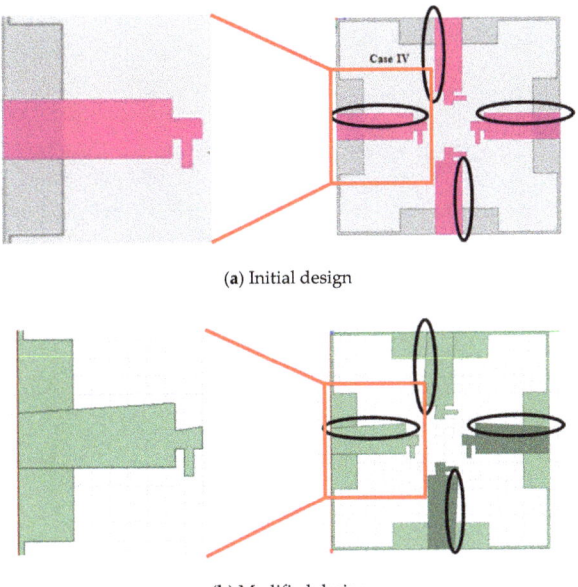

(a) Initial design

(b) Modified design

Figure 3. Antenna design for (**a**) modified and (**b**) initial antenna to match the measurement results. Area under the black eclipse represents an area of change, while the red box represents the zoomed-in area.

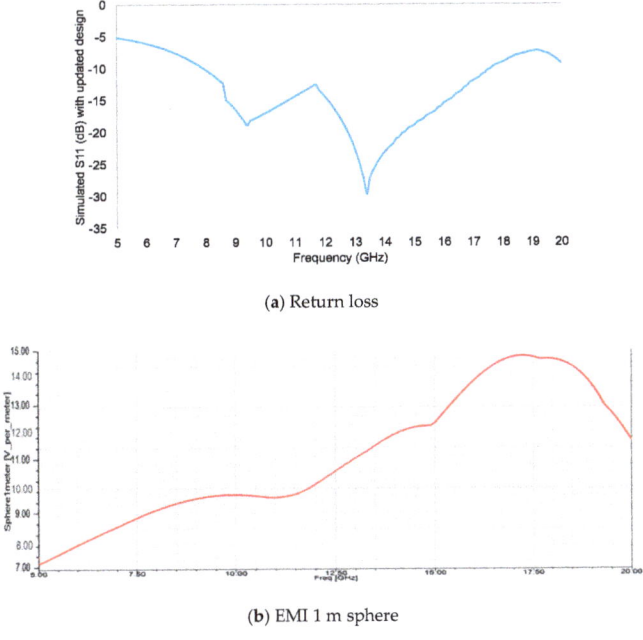

(a) Return loss

(b) EMI 1 m sphere

Figure 4. *Cont.*

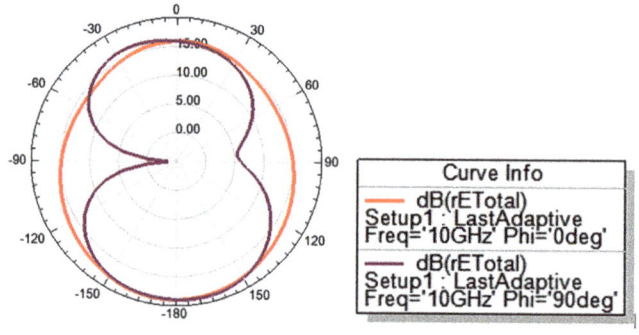

(**c**) 2-D radiation pattern

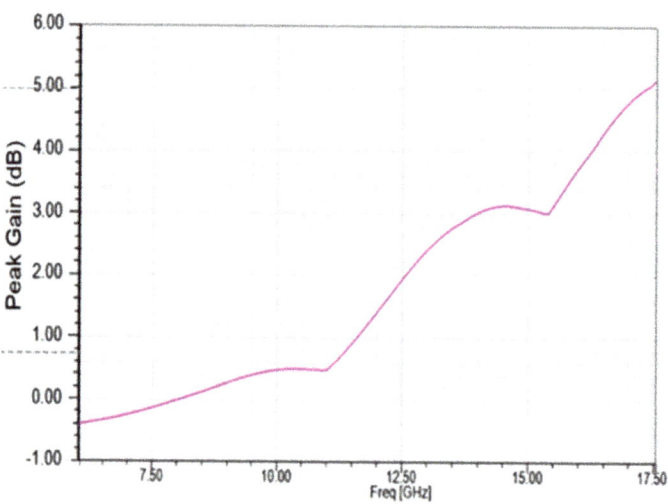

(**d**) Peak gain

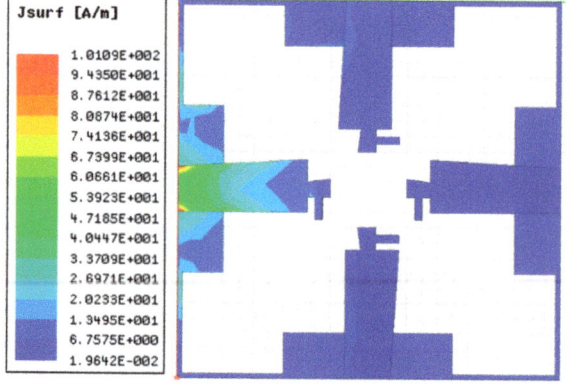

(**e**) Surface current distribution

Figure 4. *Cont.*

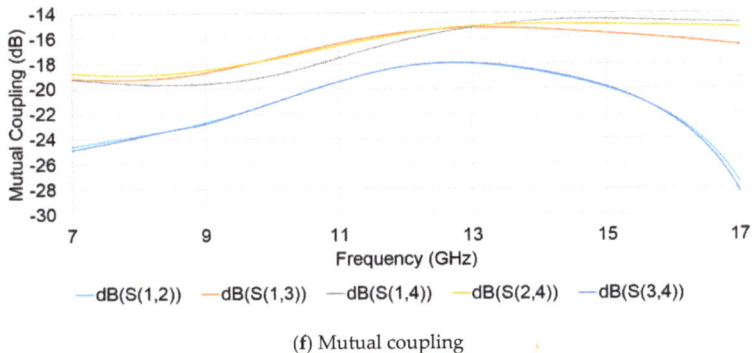

(f) Mutual coupling

Figure 4. Results for modified antenna (**a**) return loss, (**b**) electromagnetic interference (EMI), (**c**) radiation pattern, (**d**) peak gain, (**e**) surface current distribution, and (**f**) mutual coupling.

3. 2^K Factorial

Altering the ten structural parameters helps in reducing EMI, whereas changing the operating frequency affects its area of application [30]. Table 1 shows an example of the different structural parameters that can affect the return loss, bandwidth, and maximum EMI. As full factorial design requires 2^{10} experiments, which is too excessive, fractional 2^K factorial was employed [31] with a $\frac{1}{64}$ fraction, which means we can identify the important structural parameters from 32 simulation runs. The maximum and minimum values taken for each parameter are 50% and 150% of their nominal value (shown in Figure 1b). Table 2 provides respective P values for three different responses for which fractional 2^K factorials were performed. The P value helps in identifying the importance of the parameters [32].

Table 1. Examples of the changes in the antenna's structural parameters on its performance.

The Unit for the Parameters Is mm										Return Loss (dB)	Bandwidth (GHz)	Maximum EMI (V/m)
A	b	c	d	tl	tw	ga	gb	gc	gd			
0.5	1	0.75	0.75	12.75	4.5	16.5	1.5	12.5	0.45	−41.7	8.47	15.54
1.5	1	0.25	2.25	12.75	4.5	5.5	1.5	37.5	0.15	−34.2	8.51	15.78
0.5	1	0.25	0.75	4.25	1.5	5.5	1.5	37.5	0.45	−12.5	5.4	16.54

Table 2. List of significant parameters for optimization. None of the interaction terms other than these parameters were found to have P value less than 0.05, hence they are not statistically significant and are not shown here.

Parameter (mm)	P Value (Return Loss (dB))	P Value (Bandwidth (GHz))	P Value (EMI (V/m))
tl	0.009	0.027	0.054
tw	0.012	0.048	0.089
ga	0.043	0.079	0.081
gb	0.014	0.010	0.142

4. Optimization Results and Discussion

After the key structural parameters were identified from the factorial design, GA was employed for optimization. The GA optimizer setup included maximum number of generations (10,000), number of individuals for parents (30), number of mating pools per individual (30), number of individuals for children (30), number of survivors (10), and selection pressure for the next generation (10). For the reproduction setup, a uniform distribution mutation type was utilized with 0.1 uniform mutation probability, 0.5 individual mutation probability, 0.2 variable mutation probability, and 0.05 standard deviation [33]. These GA set up values were taken from [33], which also employed GA to optimize antenna design parameters with highly accurate results.

The optimized values of the parameters are shown in Table 3. Figure 5 shows the optimal design of the antenna in comparison with the initial antenna design.

Table 3. Initial and optimized values of tl, tw, ga, and gb obtained from the genetic algorithm (GA).

Parameter	Nominal Value as Initial Design (mm)	Optimized Value (mm)
tw	3.0	2.55
tl	8.5	8.62
ga	11.0	7.62
gb	3.0	3.46

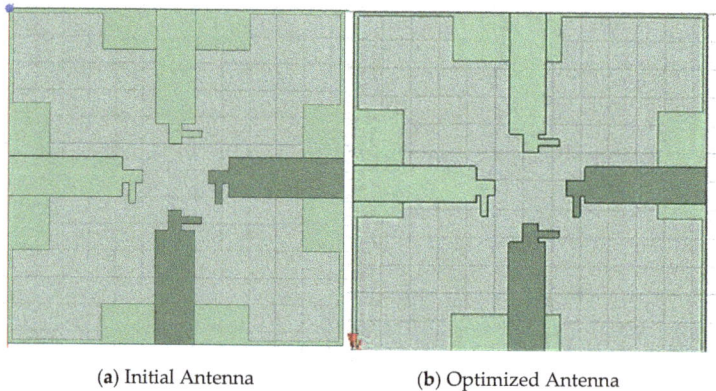

Figure 5. (a) Initial antenna and (b) optimized antenna, to show the difference in the design.

Figure 6 shows the EMI results for the nominal initial antenna design and the optimized design. We can see that in the frequency range covered by the antenna (shown in Figure 7 where the return loss is less than −10 dB), EMI was reduced by about 1 V/m. The initial antenna design covers a frequency range from 7.84 to 16.44 GHz, while the optimized antenna covers a frequency range from 7.72 to 17.25 GHz In Figure 7, the major change can be seen in the frequency at which the return loss is at a minimum.

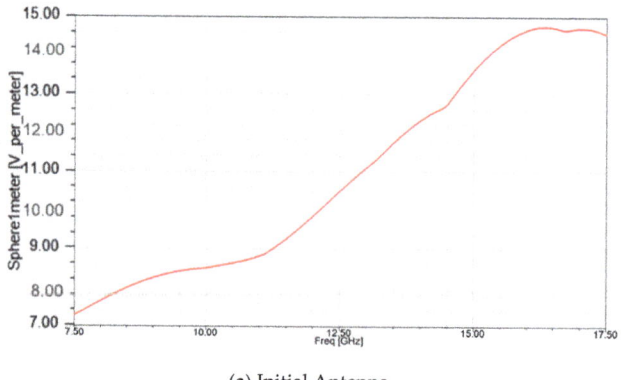

(a) Initial Antenna

Figure 6. *Cont.*

Appl. Sci. **2020**, *10*, 3117

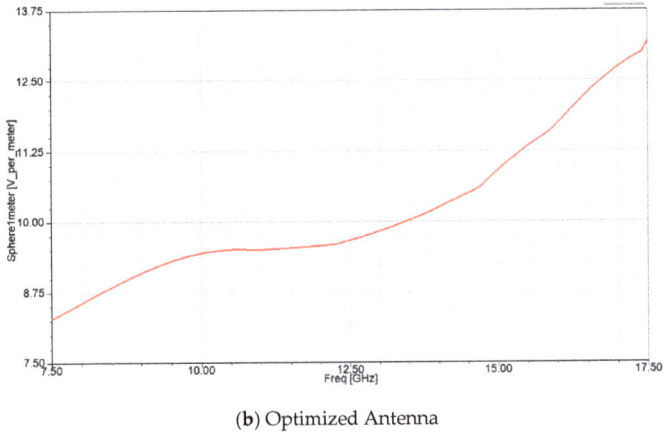

(**b**) Optimized Antenna

Figure 6. EMI for the 1 m spherical environment from the antenna.

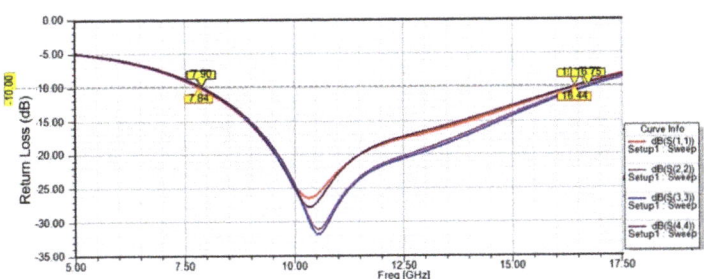

(**a**) Simulated return loss for initial antenna.

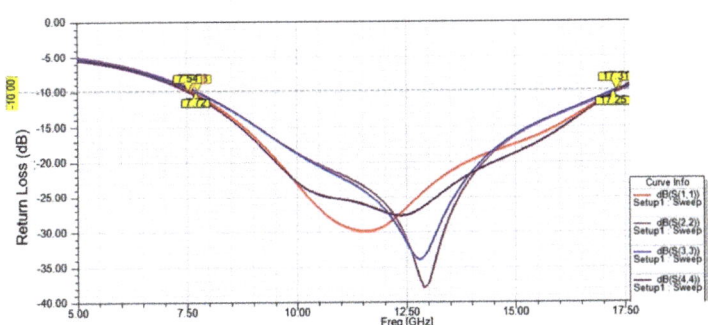

(**b**) Simulated return loss for optimized antenna.

Figure 7. Return loss of antenna.

The antenna was excited using a wave port excitation at 10.3 GHz frequency and it was observed that the maximum surface current distribution concentrated at the excited port, and this is depicted in Figure 8. Surface current distribution was reduced for the same excitation in the optimized antenna, by almost half of that of the initial antenna. Lower current density points towards lower power consumption of the antenna.

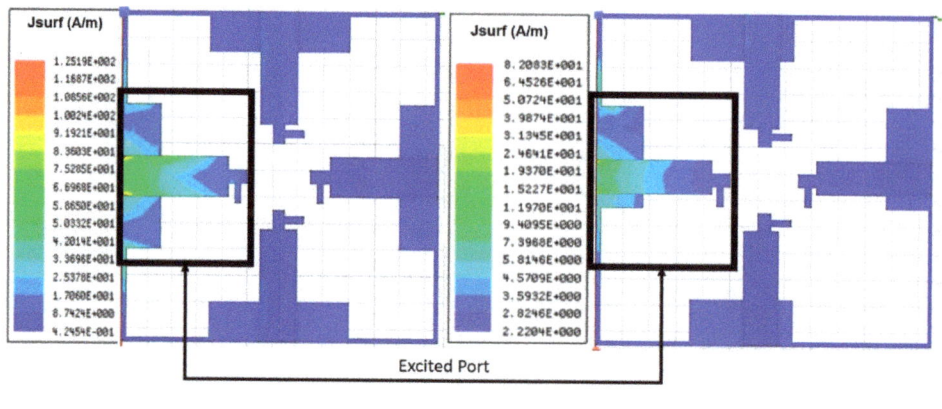

(a) Initial Antenna (b) Optimized Antenna

Figure 8. Surface current distribution.

The 2-D and 3-D radiation patterns obtained in the E-fields and H-fields of the initial and optimized antennas at 10 GHz (operating frequency) was simulated and depicted in Figure 9. The antenna was simulated in the xz-plane orientation, where red and violet color patterns in Figure 9a and b represent radiation patterns for phi angles of 0° and 90°, respectively, and the theta angle is 0°. 3-D radiation patterns help to study the spread from the antenna into the environment. It is noticed that both the initial and optimized antennas achieved omnidirectional radiation patterns at a central frequency.

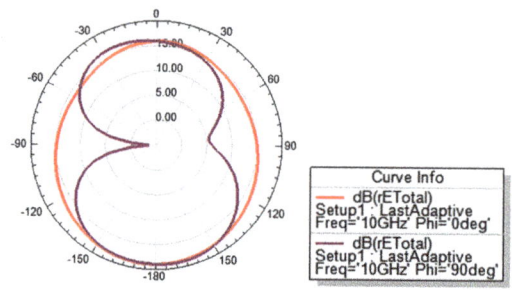

(a) 2-D radiation pattern for the initial antenna.

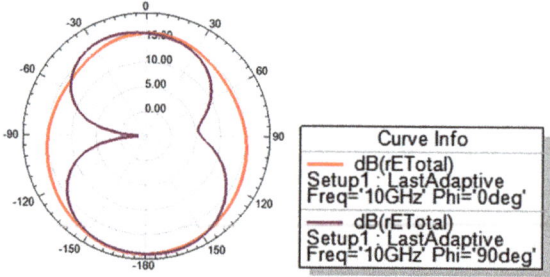

(b) 2-D radiation pattern for the optimized antenna.

Figure 9. *Cont.*

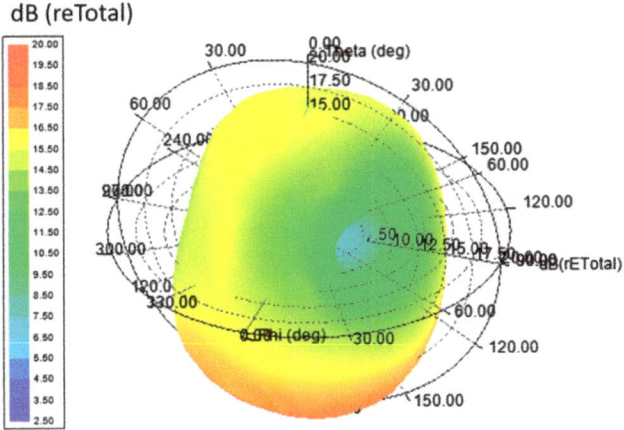

(c) 3-D radiation pattern for the initial antenna.

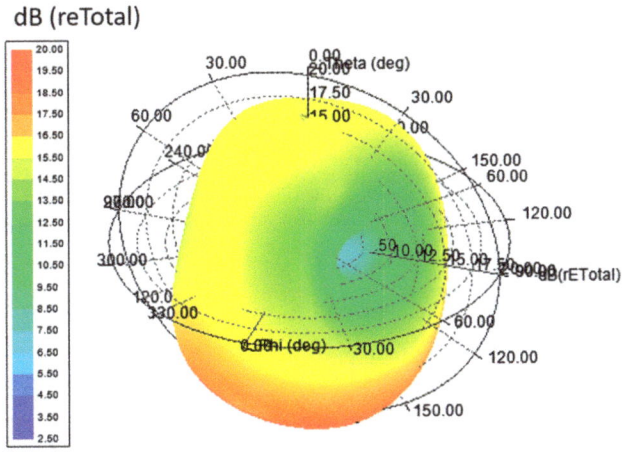

(d) 3-D radiation pattern for the optimized antenna.

Figure 9. Radiation patterns for the initial and optimized antennas: (**a**,**b**) show 2-D radiation patterns and (**c**,**d**) show 3-D radiation patterns.

Peak gain was plotted with respect to the operating frequency of the antenna as shown in Figure 10. An increase in peak gain was observed with the increase in operating frequency: 0.7 dB to 4.95 dB was the average peak gain of the initial antenna design, whereas 0 dB to 5 dB was the average peak gain for the optimized antenna for the given frequency range (where return loss was less than −10 dB). The advantage of stubs is that it helps to achieve small peak gains at lower operating frequencies [34]. A small variation of 0.05 dB found in the peak gain of an antenna is because of the reduced ground plane [34].

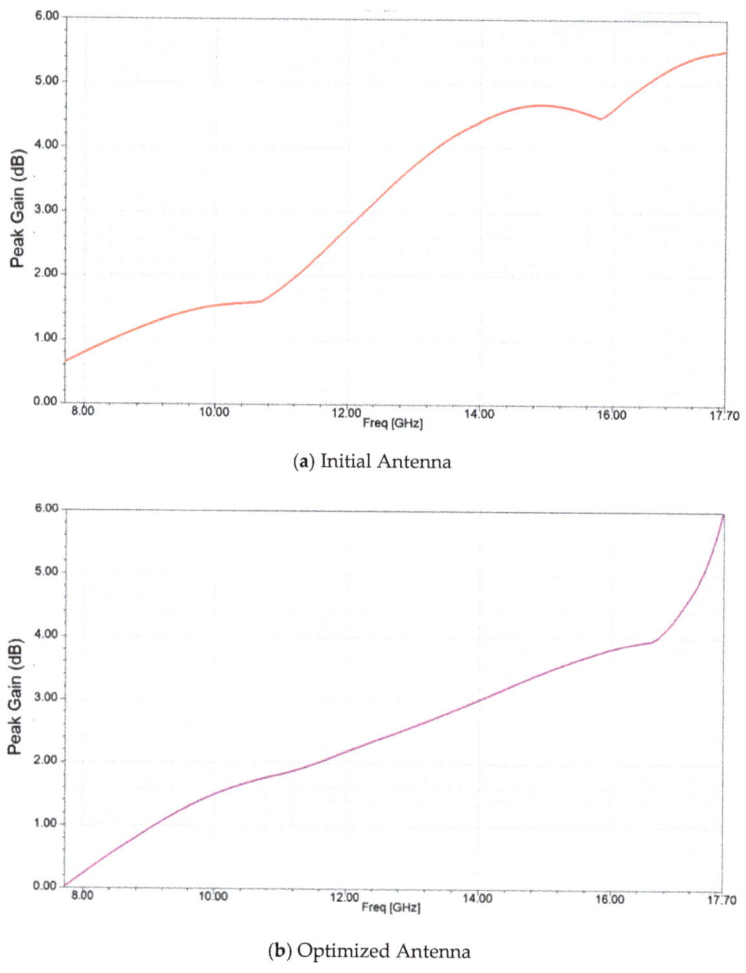

Figure 10. Peak gain in dB for initial and optimized antenna designs.

Mutual coupling between ports for the initial (found to be < −15 dB) and optimized antennas (found to be < −12 dB) are shown in Figure 11. Higher isolation was observed for ports 1 to 2 and 3 to 4 due to the identical structure of elements. The peak values of the isolation (within the obtained band) were −22 dB and −21.24 dB for the initial and optimized antennas, respectively, at 10.3 GHz. Thus, the isolation was sufficiently good with the optimized antenna. Table 4 represent the difference between initial and optimized antenna parameters.

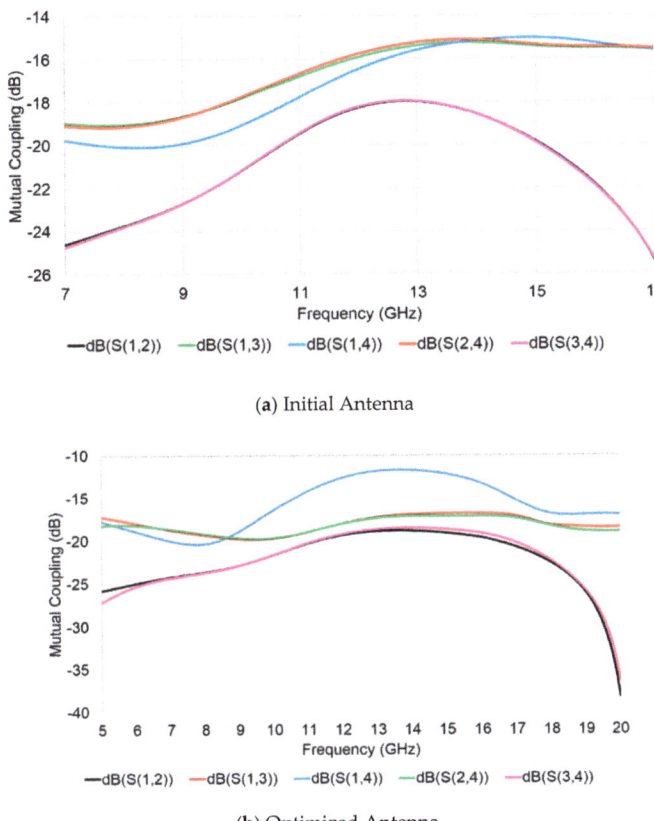

Figure 11. S-parameters of the MIMO antenna for mutual coupling between ports for (**a**) the initial antenna and (**b**) the optimized antenna.

Table 4. Antenna parameter values for the initial and optimized antennas.

Parameter	Initial Antenna Design	Optimized Antenna Design
Bandwidth (GHz)	7.84–16.44 GHz	7.72–17.25 GHz
Minimum return loss (dB)	−32	−37
Maximum peak gain (dB)	5.55	5.99
Maximum surface current density (A/m)	125.19	82.083
Maximum EMI (V/m)	14.60	13.25
Radiation intensity (dB)	15.42	15.42

5. Conclusions

In this paper, a miniaturized single element of a T-shaped MIMO antenna was designed, optimized, and simulated. The 2^K factorial and genetic algorithm optimization techniques were used for the optimization of the antenna. It was found that gb, ga, tl, and tw, as defined in Figure 1, were the major dimensional parameters affecting the return loss, peak gain, bandwidth, radiation pattern, surface current distribution, and EMI of the antenna. The optimized antenna had a wider frequency band that ranged from 7.72 to 17.25 GHz. This corresponded to a 1 GHz improvement in the bandwidth due to the improved return loss that provides new opportunities for the antenna's utilization in different applications. The optimized antenna had lower current distribution that gives lower power dissipation and its 1 m sphere EMI was also reduced.

Author Contributions: Conceptualization, C.M.T.; methodology, D.K.; software, D.K., V.S., V.P., N.T.; validation, D.K.; formal analysis, D.K., V.S., V.P.; investigation, D.K.; resources, C.M.T.; data curation, D.K., V.S., V.P.; writing—original draft preparation, D.K.; writing—review and editing, C.M.T.; visualization, D.K., N.T.; supervision, C.M.T.; project administration, C.M.T.; funding acquisition, C.M.T. All authors have read and agreed to the published version of the manuscript.

Funding: This work was funded by the Chang Gung University research Grants QZRPD123 and CIRPD2F0024.

Conflicts of Interest: The authors declare no conflict of interest.

References

1. Kragalott, M.; Kluskens, M.S.; Zolnick, D.A.; Dorsey, W.M.; Valenzi, J.A. A toolset independent hybrid method for calculating antenna coupling. *IEEE Trans. Antennas Propag.* **2011**, *59*, 443–451. [CrossRef]
2. Saxena, S.; Kanaujia, B.K.; Dwari, S.; Kumar, S.; Tiwari, R. A Compact Dual-Polarized MIMO Antenna with Distinct Diversity Performance for UWB Applications. *IEEE Antennas Wirel. Propag. Lett.* **2017**, *16*, 3096–3099. [CrossRef]
3. Rajkumar, S.; Vivek Sivaraman, N.; Murali, S.; Selvan, K.T. Heptaband swastik arm antenna for MIMO applications. *IET Microw. Antennas Propag.* **2017**, *11*, 1255–1261. [CrossRef]
4. Roshna, T.K.; Deepak, U.; Sajitha, V.R.; Vasudevan, K.; Mohanan, P. A compact UWB MIMO antenna with reflector to enhance isolation. *IEEE Trans. Antennas Propag.* **2015**, *63*, 1873–1877. [CrossRef]
5. Krishna, R.V.S.R.; Kumar, R. A Dual-Polarized Square-Ring Slot Antenna for UWB, Imaging, and Radar Applications. *IEEE Antennas Wirel. Propag. Lett.* **2016**, *15*, 195–198. [CrossRef]
6. Jafri, S.I.; Saleem, R.; Shafique, M.F.; Brown, A.K. Compact reconfigurable multiple-input-multiple-output antenna for ultra wideband applications. *IET Microw. Antennas Propag.* **2016**, *10*, 413–419. [CrossRef]
7. Khan, A.A.; Jamaluddin, M.H.; Aqeel, S.; Nasir, J.; Kazim, J.U.R.; Owais, O. Dual-band MIMO dielectric resonator antenna for WiMAX/WLAN applications. *IET Microw. Antennas Propag.* **2017**, *11*, 113–120. [CrossRef]
8. Iqbal, A.; Saraereh, O.A.; Ahmad, A.W.; Bashir, S. Mutual Coupling Reduction Using F-Shaped Stubs in UWB-MIMO Antenna. *IEEE Access* **2017**, *6*, 2755–2759. [CrossRef]
9. Tian, R.; Lau, B.K.; Ying, Z. Multiplexing efficiency of MIMO antennas. *IEEE Antennas Wirel. Propag. Lett.* **2011**, *10*, 183–186. [CrossRef]
10. Liu, X.; Bialkowski, M.E. Effect of antenna mutual coupling on MIMO channel estimation and capacity. *Int. J. Antennas Propag.* **2010**, *2010*, 1–9. [CrossRef]
11. Abdul Haleem, M. On the capacity and transmission techniques of massive MIMO Systems. *Wirel. Commun. Mob. Comput.* **2018**, *2018*, 1–9. [CrossRef]
12. Khalighi, M.A.; Raoof, K.; Jourdain, G. Capacity of wireless communication systems employing antenna arrays, a tutorial study. *Wirel. Pers. Commun.* **2002**, *23*, 321–352. [CrossRef]
13. Du, J.; Li, Y. Optimization of antenna configuration for MIMO systems. *IEEE Trans. Commun.* **2005**, *53*, 1451–1454. [CrossRef]
14. Kiehbadroudinezhad, S.; Noordin, N.K.; Sali, A.; Abidin, Z.Z. Optimization of an antenna array using genetic algorithms. *Astron. J.* **2014**, *147*, 1–13. [CrossRef]
15. Telzhensky, N.; Leviatan, Y. Novel method of UWB antenna optimization for specified input signal forms by means of genetic algorithm. *IEEE Trans. Antennas Propag.* **2006**, *54*, 2216–2225. [CrossRef]
16. Binitha, S.; Siva Sathya, S. A survey of bio inspired optimization algorithms. *Int. J. Soft Comput. Eng.* **2012**, *2*, 137–151.
17. Hussein, A.H.; Abdullah, H.H.; Salem, A.M.; Khamis, S.; Nasr, M. Optimum design of linear antenna arrays using a hybrid MoM/GA algorithm. *IEEE Antennas Wirel. Propag. Lett.* **2011**, *10*, 1232–1235. [CrossRef]
18. Zhou, D.; Abd-Alhameed, R.A.; See, C.H.; Bin-Melha, M.S.; Zainal-Abdin, Z.B.; Excell, P.S. New antenna designs for wideband harmonic suppression using adaptive surface meshing and genetic algorithms. *IET Microw. Antennas Propag.* **2011**, *5*, 1054–1061. [CrossRef]
19. Robinson, J.; Rahmat-Samii, Y. Particle swarm optimization in electromagnetics. *IEEE Trans. Antennas Propag.* **2004**, *52*, 397–407. [CrossRef]

20. Mohammed, H.J.; Abdullah, A.S.; Ali, R.S.; Abd-Alhameed, R.A.; Abdulraheem, Y.I.; Noras, J.M. Design of a uniplanar printed triple band-rejected ultra-wideband antenna using particle swarm optimisation and the firefly algorithm. *IET Microw. Antennas Propag.* **2016**, *10*, 31–37. [CrossRef]
21. Johnson, J.M.; Rahmat-Samii, Y. Genetic algorithms in engineering electromagnetics. *IEEE Antennas Propag. Mag.* **1997**, *39*, 7–21. [CrossRef]
22. Panduro, M.A.; Brizuela, C.A. A comparative analysis of the performance of GA, PSO and DE for circular antenna arrays. In Proceedings of the IEEE Antennas and Propagation Society International Symposium, Charleston, SC, USA, 1–5 June 2009; IEEE: Piscataway, NJ, USA, 2009; pp. 1–4.
23. Binelo, M.O.; De Almeida, A.L.F.; Cavalcanti, F.R.P. MIMO array capacity optimization using a genetic algorithm. *IEEE Trans. Veh. Technol.* **2011**, *60*, 2471–2481. [CrossRef]
24. Oprime, P.C.; Pureza, V.M.M.; De Oliveira, S.C. Systematic sequencing of factorial experiments as an alternative to the random order. *Gest. Prod.* **2017**, *24*, 108–122. [CrossRef]
25. Cavalcanti, F.R.P. *Resource Allocation and MIMO for 4G and Beyond*; Springer: New York, NY, USA, 2014.
26. E Silva Neto, A.S.; de Macedo Dantas, M.L.; dos Santos Silva, J.; César Chaves Fernandes, H. Antenna for fifth generation (5G) using a EBG structure. In *New Contributions in Information Systems and Technologies*; Springer: Cham, Switzerland, 2015; pp. 33–38.
27. Hoang, T.V.; Le, T.T.; Li, Q.Y.; Park, H.C. Quad-Band Circularly Polarized Antenna for 2.4/5.3/5.8-GHz WLAN and 3.5-GHz WiMAX Applications. *IEEE Antennas Propag. Lett.* **2016**, *15*, 1032–1035. [CrossRef]
28. Liu, X.L.; Wang, Z.D.; Yin, Y.Z.; Ren, J.; Wu, J.J. A compact ultrawideband MIMO antenna using QSCA for high isolation. *IEEE Antennas Wirel. Propag. Lett.* **2014**, *13*, 1497–1500. [CrossRef]
29. Tang, T.C.; Lin, K.H. An ultrawideband MIMO antenna with dual band-notched function. *IEEE Antennas Wirel. Propag. Lett.* **2014**, *13*, 1076–1079. [CrossRef]
30. Kishk, A. *Advancement in Microstrip Antennas with Recent Applications*; InTech: Rijeka, Croatia, 2013.
31. Hilow, H. Minimum cost linear trend free fractional factorial designs. *J. Stat. Theory Pract.* **2012**, *6*, 580–589. [CrossRef]
32. Dahiru, T. P-Value, a true test of statistical significance? A cautionary note. *Ann. Ibadan Postgrad. Med.* **2008**, *6*, 21–26. [CrossRef]
33. Mohammed, H.J.; Abdulsalam, F.; Abdulla, A.S.; Ali, R.S.; Abd-Alhameed, R.A.; Noras, J.M.; Abdulraheem, Y.I.; Ali, A.; Rodriguez, J.; Abdalla, A.M. Evaluation of genetic algorithms, particle swarm optimisation, and firefly algorithms in antenna design. In Proceedings of the 2016 13th International Conference on Synthesis, Modeling, Analysis and Simulation Methods and Applications to Circuit Design, SMACD 2016, Lisbon, Portugal, 27–30 June 2016; IEEE: Piscataway, NJ, USA, 2016; pp. 1–4.
34. John, M.; Evans, J.A.; Ammann, M.J.; Modro, J.C.; Chen, Z.N. Reduction of ground-plane-dependent effects on microstrip-fed printed rectangular monopoles. *IET Microw. Antennas Propag.* **2008**, *2*, 42–47. [CrossRef]

© 2020 by the authors. Licensee MDPI, Basel, Switzerland. This article is an open access article distributed under the terms and conditions of the Creative Commons Attribution (CC BY) license (http://creativecommons.org/licenses/by/4.0/).

Article

Investigate the Equivalence of Neutrons and Protons in Single Event Effects Testing: A Geant4 Study

Yueh Chiang [1,2], Cher Ming Tan [3,4,5,6,*], Tsi-Chian Chao [1,7], Chung-Chi Lee [1,7] and Chuan-Jong Tung [7]

1. Department of Medical Imaging and Radiological Sciences, College of Medicine, Chang Gung University, Tao-Yuan 333, Taiwan; D0503203@cgu.edu.tw (Y.C.); chaot@mail.cgu.edu.tw (T.-C.C.); cclee@mail.cgu.edu.tw (C.-C.L.)
2. Department of Radiation Oncology, Chang Gung Memorial Hospital, Tao-Yuan 333, Taiwan
3. Center for Reliability Sciences and Technologies, Chang Gung University, Tao-Yuan 333, Taiwan
4. Urology Department, Chang Gung Memorial Hospital, Tao-Yuan 333, Taiwan
5. Center for reliability engineering, Ming Chi University of Technology, New Taipei City 243, Taiwan
6. Department of electronic engineering, College of Engineering, Chang Gung University, Tao-Yuan 333, Taiwan
7. Particle Physics and Beam Delivery Core Laboratory, Institute for Radiological Research, Chang Gung University/Chang Gung Memorial Hospital, Tao-Yuan 333, Taiwan; cjtung@mail.cgu.edu.tw
* Correspondence: cmtan@cgu.edu.tw

Received: 15 March 2020; Accepted: 2 May 2020; Published: 6 May 2020

Abstract: Neutron radiation on advanced integrated circuits (ICs) is becoming important for their reliable operation. However, a neutron test on ICs is expensive and time-consuming. In this work, we employ Monte Carlo simulation to examine if a proton test can replace or even accelerate the neutron test, and we found that 200 MeV protons are the closest to resembling neutron radiation with five main differences. This 200 MeV concur with the suggestion from National Aeronautics and Space Administration (NASA, Washington, DC, USA). However, the impacts of the five differences on single event effects (SEEs) require future work for examination.

Keywords: single event effects; linear energy transfer; Monte Carlo simulation; radiation hardness

1. Introduction

Technological developments bring smaller and faster devices in integrated circuits that operate at reduced bias voltages. However, they also suffer from increased susceptibility to neutrons. These neutrons can cause single event effects (SEEs) on the integrated circuits, rendering their temporary loss of function. Such temporary loss of function may be a critical issue in many applications, especially for implanted medical electronics such as pacemakers [1].

Consequently, radiation tests are becoming necessary to ensure reliable applications of these circuits, especially in applications where their exposure to radiation intensity might be higher. An example of radiation test can be found in Intel. Seifert et al. reported their radiation test results in Intel, demonstrating that radiation-induced soft error rate (SER) improvements in the 14 nm generation high-k+ metal gate as compared to the bulk tri-gate technology [2]. Its taller and narrower structure minimized the charge collection owing to a smaller, sensitive volume. There are many more reported radiations tests as can be seen in the annual workshop on soft error Silicon Errors in Logic—System Effects (SELSE). The SELSE workshop provides a forum for discussion of current research and practice in system-level error management. Participants from industry and academia explore both current technologies and future research directions (including nanotechnology). SELSE is soliciting papers that address the system-level effects of errors from a variety of perspectives: architectural, logical, circuit-level, and semiconductor processes where several companies are reporting their radiation tests in the workshop.

However, a fast neutron test that resembles normal operating conditions is expensive due to the long test duration, and the test can only be done in limited facilities such as TRIUMF (Vancouver, BC, Canada; up to 400 MeV neutrons), Los Alamos Neutron Science Center (LANSCE, Los Alamos, NM, USA; up to 750 MeV neutrons), and ISIS Neutron and Muon Source (Oxford, UK; up to 400 MeV neutrons). Proton facilities, however, are easier to access worldwide. Another advantage of using protons to replace neutrons for SEE testing is that the protons are charged particles which can be easily accelerated and focused. Wei et al. showed the possibility and challenge of using a medical proton facility to do SEE testing [3]. In fact, the use of protons to study the radiation effect on electronics in various radiation environments has been practiced. O'Neill et al. proposed that 200 MeV proton can be used to mimic the radiation environment at low Earth orbit (LEO) [4], and National Aeronautics and Space Administration (NASA) used 200 MeV proton to perform their SEE tests, and successfully screened thousands of electronic parts at Indiana University Cyclotron Facility (IUCF, Bloomington, IN, USA) even though the previously reported SEE failures were due to heavy ions [4].

The SER in integrated circuit can be estimated by dose (or energy deposited) convoluted with energy-specific linear energy transfer (LET) [5]. In a radiation hardness test, the dose can be controllable by flux, but the LET is the characteristic of particles with specific energy. LET expresses the characteristics of a particle's path through materials. It is defined as the energy being transferred to a material by an ionizing particle as a function of distance and material density, in units of MeV-cm^2/mg. Dodd et al. showed that LET is the key index of SER in high-speed digital logic integrated circuits (ICs) [6]. Bagatin et al. also showed the correlation of LET and single event upset (SEU) in floating gate cells [7]. In addition to LET, the specific secondary particles can also be important because it can create different kinds of defect in the materials in integrated circuits [3].

In this work, we examine the possible use of a medical proton test facility to replace the fast neutron test at sea level using the Monte Carlo simulation known as Geant4. To evaluate the equivalence of neutron and proton SEE tests, LET of both primary and secondary particles and secondary particle yields are examined. The secondary particle yields are also important in SEE evaluation because the secondary particles, especially for heavy ions, may implant in the silicon crystal and change the electronic properties.

Subsequent quantitative verification will be performed using our newly constructed proton center at Chang Gung Memorial Hospital (Linkou District, New Taipei, Taiwan) which is a medical cyclotron (Sumitomo Heavy Industry, Tokyo, Japan) with a maximum beam current around 300 nA at 230 MeV. The energy spread is less than 10% for low energy (30 MeV) and 1% for higher energy (110 + MeV).

2. Materials and Methods

In this work, we focus on silicon-based semiconductors. However, there are always back-end interconnected structures that contain metals with high Z materials, such as copper and refractory metals. As the spallation cross-sections of neutrons and protons are correlated to Z, the high LET secondary particles are mostly generated in the back-end structures after neutron and proton radiation, and these particles can hit the semiconductor region producing SEEs.

Another common silicon-based semiconductor material is silicon-germanium (SiGe). SiGe is an upcoming advanced silicon-based IC technology as SiGe technology effectively merges the desirable attributes of conventional silicon-based CMOS manufacturing (high integration levels, at high yield and low cost) with the extreme levels of transistor performance attainable in classical III–V heterojunction bipolar transistors (HBTs) through bandgap engineering. This renders SiGe HBTs with several key merits with respect to operation across a wide variety of so-called "extreme environments", potentially with little or no process modification, ultimately providing compelling advantages at the circuit and system level, and across a wide class of envisioned commercial and defense applications. Thus, both silicon and SiGe materials are studied in this work.

Since back-end interconnected structures are the major sources of high LET secondary particles that can induce SEEs, we need to confirm the hypothesis that the spectra of secondary particles generated from the back-end structures by protons are similar to those from the fast neutrons in the LANSCE, as shown in Figure 1 [8]. We performed this hypothesis testing using Geant4 Monte Carlo simulations [9].

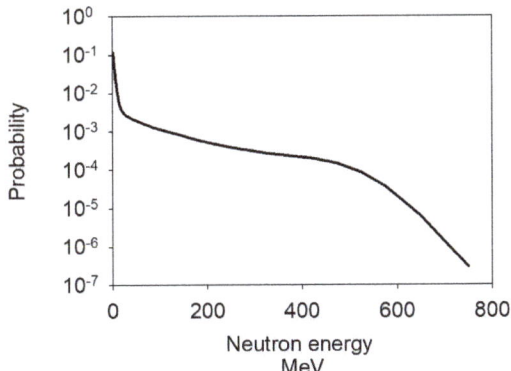

Figure 1. The Los Alamos Neutron Science Center (LANSCE) broad band neutron spectrum used in this study [8].

Geant4 is a well benchmarked general-purpose Monte Carlo code for both macroscopic and microelectronic scales. Intel has used Geant4 to build their "Intel Radiation Tool" for radiation effect simulation [10]. Weller et al. also established a Monte Carlo approach for estimating SEEs using Geant4 with TCAD [11].

2.1. Monte Carlo Simulation

To simulate the LETs and secondary particle yields in semiconductor devices, Geant4 10.04.p02 was used in this work [12]. The physics list used in this study is QGSP_BIC_HP_EM4 with radioactive decay enabled. In particular, G4EmStandardPhysics_option4 was implemented for modeling Electromagnetic process, G4HadronElasticPhysics for elastic process of hadrons, G4HadronPhysicsQGSP_BIC for inelastic process of hadrons, G4RadioactiveDecayPhysics for radioactive decay, and G4IonBinaryCascadePhysics for inelastic process. The quark-gluon string precompound (QGSP) model was implemented to handle collision of high-energy hadrons, and the binary cascade model was used for inelastic process of hadrons. The high precision data were set for the low-energy neutron and light ions. The details of the hadron interaction and ionization model in Geant4 can be found in Truscott et al. [13].

Since elastic and inelastic interactions due to the nuclear reactions of radiation particles and materials are critical to predict secondary particle yields, the Joint Evaluated Fission and Fusion File (JEFF) 3.3 Nuclear Data Library was added to the neutron simulation, and the G4TENDL data set was also added for high precision particle transportation.

To obtain more accurate secondary yields, the nuclear decay model was enabled with the function DO_NOT_ADJUST_FINAL_STATE. This is a function in Geant4 which instructs the simulation software not to generate artificial gamma rays for the purpose of satisfying the energy and momentum conservation in some nuclear reactions. The simulation will follow the ENDF-6 libraries [14]. Moreover, the cut-off range for secondary particles was set to 0.01 nm, so that all the secondary particles can continuously slow down to an approximation range longer than 0.01 nm [15].

The studied primary incident particles are the LANSCE broad band neutron and protons with selected energies. These particles were generated using the Geant4 general particle source (GPS). For neutrons, a spectrum from LANSCE was converted to a probability distribution (Figure 1) and inputted to the GPS. For protons, eight monoenergetic protons of 10, 30, 50, 63, 105, 150, 200, and 230 MeV respectively were generated, and only one of the above energies was simulated at a time. In each simulation case, the number of histories was 10^9. History in the context of Monte Carlo simulation refers to a record of a primary particle from being generated to being stopped.

All the primary particles were generated on the top of the structure and transported by the Geant4 process class with the physics list mentioned above. The secondary particle species generated due to the interactions between the primary particles and materials was recorded by the Geant4 stepping class when the particles entered the detection layer. To prevent the partial volume effect, which usually happens when the scoring volume overlaps two or more materials at the geometry boundary [16], the energy deposition was calculated by the Geant4 tracking class which sums up all the energy deposition in each step from the track passing through the detection layer. The details of the Monte Carlo technique used in this work and the associated issues can be found in a study by Chiang and colleagues [17].

Since there is no method to measure the LET spectrum in an actual integrated circuit, our simulation is benchmarked with the simulation results of O'Neill for pure silicon [16], and good agreement was obtained. In fact, the same simulation setup has been used for simulating the microdosimetry property of protons and photons in biological targets with 1 µm diameter and again good agreement was obtained. This work was reported by Hsing et al. [18].

2.2. Material Structure

The material examined in this work is a back-end structure modified from Zhang et al. [12]. The aluminum layer is replaced by copper and the detection layer is 100 nm thick, as shown in Figure 2a. No titanium layer is included in the simulation of this work. Figure 2b shows the same structure but with an additional thin SiGe layer to evaluate the effect of SiGe on SEE events due to neutron and proton radiation. Other metallization structures will be examined in our other work.

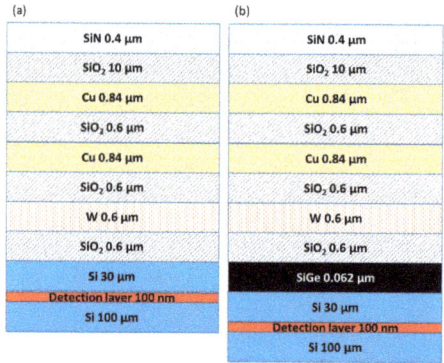

Figure 2. The layer structure (**a**) without SiGe and (**b**) with SiGe used in this simulation (not to scale).

To achieve a charged-particle equilibrium (CPE) and maintain simulation efficiency, the diameter of the structure is set to 1 mm, which is much larger than the secondary particle range. Monte Carlo simulation may have boundary crossing problems when particles transport from a large volume to a small volume. To minimize this effect, the step size in this study is limited by a stepping function that each step cannot be longer than 0.01% of its calculated range. The final step cannot be bigger than 0.1 nm. In addition, the skin parameter in this study is set to three, which means that single scattering mode will activate three elastic mean free paths before the boundary.

2.3. Data Analysis

LET is calculated from the energy deposited in the detection layer divided by its thickness (100 nm) and the density of silicon (2330 mg/cm^3). The energy deposited is calculated by the sum of the energy imparted by all the events from all the tracks passed through the detection layer.

To compare the similarity of each LET curve or compare the similarity of the secondary particle yields, an evaluation index (EI) is defined for ith LET bins or i-types of secondary particles, as given in Equation (1). In the EI definitions, LET_i is the differential fluence in the ith LET bin and Y_i is the secondary particle yield at the test condition. $A_{ref,i}$ is the corresponding quantity at the reference condition, which is simulated from the LANSCE broad band neutron spectrum. To prevent dividing by zero, any $A_{test,\,i}$ with $A_{ref,i} = 0$ is ignored. In LET cases, the bin size for analysis was 0.2 MeV-cm^2/mg and the counts were analyzed using a log scale.

EI is defined as the root mean square of relative error as follows:

$$EI = \sqrt{\frac{\sum_{i=1}^{N}\left[\frac{(A_{test,\,i} - A_{ref,i})}{A_{ref,i}}\right]^2}{N}} \quad (1)$$

here A_i is either LET_i or Y_i. The smaller EI will indicate better equivalency between two sets of data. The EI will be zero if the test protons generate identical LET differential fluences compared to those generated by the LANSCE neutron, that is, $(A_{test,\,i} - A_{ref,i}) = 0$, the same for secondary particle yield comparisons.

3. Results and Discussions

To compare the equivalence of neutrons and protons in SEE testing for the layer structure. LET, secondary yield, and energy deposited are evaluated in this work. Additionally, the effects of SiGe are also considered.

3.1. LET Difference between Neutrons and Protons

To examine if protons can be used to replace neutrons for SEE testing, the most important consideration is the equivalence of the LET spectrum as LET is a key determining parameter for the SEE. Figure 3 shows the LET spectra of the structure without SiGe being irradiated by protons and neutrons.

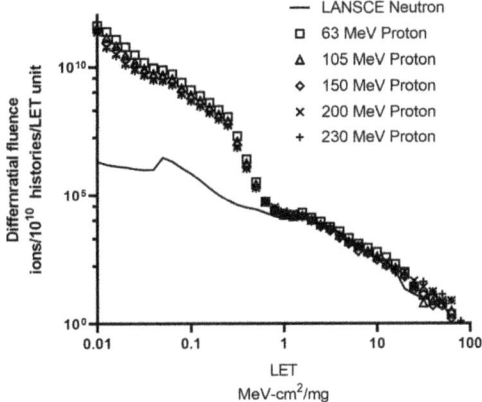

Figure 3. Linear energy transfer (LET) spectra in a structure without silicon-germanium (SiGe) irradiated by 63, 105, 150, 200, and 230 MeV proton and LANSCE neutron.

The LET spectra of the examined geometry irradiated with LANSCE neutron and selected monoenergetic protons were plotted in Figure 3. The largest visible difference in the LET spectra is below 1 MeV-cm^2/mg, where protons generated 1000 times more events than neutrons did. Due to the low LET, this visible difference has minimal impact on single event effects, whilst it could cause the total ionization dose (TID) effects under prolonged radiation exposure. However, the cross section of the SEE can be altered by TID as described by Schwank et al. [19] and Lorne et al. [20].

We further compared the LET spectra of 200 MeV proton and the LANSCE neutron for LET larger than 1 MeV-cm^2/mg. The differential fluence was in good agreement in LETs between 1 to 10 MeV-cm^2/mg but was slightly deviated in LETs larger than 10 MeV-cm^2/mg. This phenomenon can be explained with the help of Figure 4, in which the LET was plotted for several major secondary particles. In parentheses, the first symbol represents incident particles and the second symbol represents particles that contribute to the LET. For example, (n, He) means that the helium is generated by the LANSCE neutron.

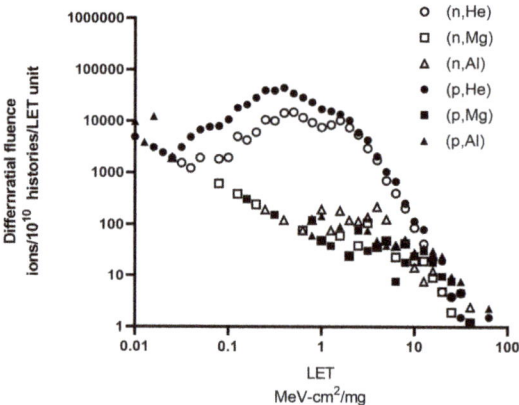

Figure 4. The LET contribution from He, Mg, and Al generated by the 200 MeV proton and the LANSCE neutron. In parentheses, the first symbol represents incident particles and the second symbol represents particles that contribute to the LET.

In the LETs between 1 to 10 MeV-cm^2/mg, events were mainly caused by helium ions, which were mostly due to the elastic interactions of high-energy particles, regardless of whether they were neutrons or protons. For LETs more than 10 MeV-cm^2/mg, 200 MeV protons gave a higher differential fluence than the LANSCE neutron because protons generate more heavy secondary ions such as aluminum and magnesium. The LANSCE neutron produces very few of these secondary particles because most of the LANSCE neutron shown in Figure 1 is less than 10 MeV which is below the threshold energy of these nuclear interactions.

The LET spectra from the studied structure irradiated by lower proton energy are rarely used, but we included them in our work (see Figure 5). In these low-energy proton irradiations, the differential fluence in LET is 10,000 times higher than neutrons for LET lower than 0.5 MeV-cm^2/mg and five times higher for LETs between 1 to 10 MeV-cm^2/mg. On the other hand, for LETs larger than 10 MeV-cm^2/mg, low-energy protons give less events. Therefore, from the above results, our further evaluation will focus on the protons with energy higher than 63 MeV.

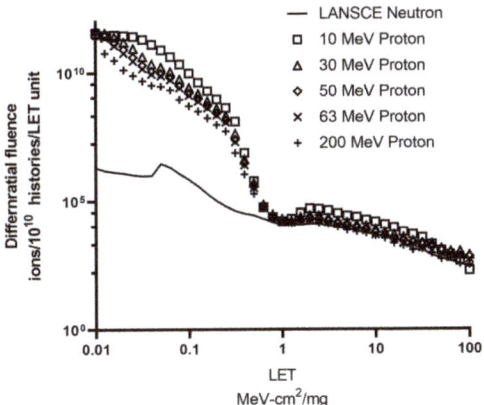

Figure 5. LET spectra in a structure without SiGe from 10, 30, 50, 63, and 200 MeV protons and LANSCE neutron.

In comparison to the results from Hiemstra et al. [21], the LET distribution in our work was much wider because the simulation of all secondary particles was performed in our study, which means that both the ionization energy loss and the nuclear interactions were simulated. Therefore, our simulations are expected to be more accurate, but it is time-consuming (takes 2–3 days per energy per condition in a computer with Intel I9-9900k CPU working at 4600 MHz and two dual channels 8 GB RAM working at 3000 MHz). The uncertainty in our simulation for the LET spectra shown in Figures 3 and 4, is quantified using the coefficient of variation (C_v). We found that the C_v is less than 10% for LETs lower than 6 MeV-cm^2/mg. However, if the LET is greater than 40 MeV-cm^2/mg, C_v is between 30% and 70%, due to the small number of events with these LETs. To be noted, compared pair with high C_v may contribute more to the EI calculation, so that the LETs larger than 6 MeV-cm^2/mg will dominate our conclusion about beam equivalency.

Another comparison is with the work from Turflinger et al. [22]; our results and theirs agree well for LETs lower than 15 MeV-cm^2/mg. For higher LETs, no comparison can be made because the work of Turflinger et al. has a 5 µm Pb/Au layer which is not present in this study.

The EI evaluation of the LET spectra is shown in Table 1, which shows that the 200 MeV proton has a LET spectrum with the best equivalence compared to the LANSCE neutron. The EI decreases with increasing proton energy, reaching the minimum at 200 MeV and increased at 230 MeV, with the exception for the 150 MeV proton, which may be questionable owing to the presence of a switch point for the hadron interaction around 150 MeV [14] in Geant4. The results shown in Table 1 are consistent with Figure 5 where 63 MeV proton gives higher differential fluence than the neutron, and 200 MeV proton gives lower differential fluence than the neutron in LETs < 10 MeV-cm^2/mg. Of course, it is also possible that the best proton to replace the LANSCE neutron is with an energy between 105 and 200 MeV. Alternatively, a mix of different proton energies could also provide a better equivalence. All these possibilities are explored later.

Table 1. Evaluation index (EI) for LET in layer structure without SiGe.

	LANSCE Neutron	63 MeV Proton	105 MeV Proton	150 MeV Proton	200 MeV Proton	230 MeV Proton
EI	0	0.290	0.274	0.296	0.250 *	0.285

* Indicates the best choice.

3.2. Secondary Particle Yield Difference between Neutron and Proton

The secondary particle species can not only produce a different LET but also change the semiconductor properties. In this study, the secondary particles with Z = 2–80 were analyzed. Figure 6 shows the secondary particle yields in structures without SiGe when it is irradiated by 63, 105, 150, 200, and 230 MeV protons and LANSCE neutron, respectively.

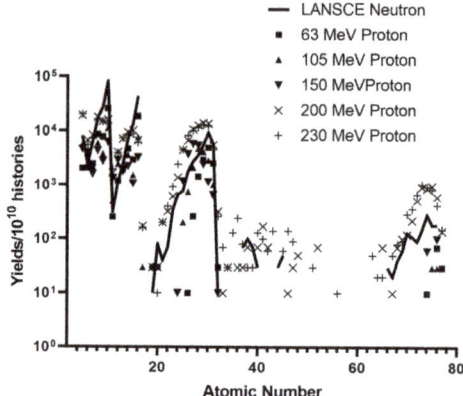

Figure 6. The secondary particle yields in structure without SiGe irradiated by 63, 105, 150, 200, and 230 MeV protons and LANSCE neutron.

In Figure 6, five groups can clearly be identified, namely alpha group, O group, Si group, Cu group, and W group, from left to right. Between copper and tungsten groups, there are some events from the cleavage fragments of W. In the secondary particle yields, the highest peak is at Z = 2, which is helium or alpha particles resulting from elastic interaction. The next two peaks, at Z = 6 and 14, are from SiO_2.

It can be seen in Figure 6 that the secondary particles distribution of the neutrons is narrower than that of the protons. The reason is that the energy of most neutrons is too low to have a spallation event. Another difference is that for Z > 14, 200 MeV protons generate more secondary particles than neutrons, but for 8 < Z < 14, neutrons give more events. The difference is mostly from the last layer of the structure, which is SiO_2. In the Z = ~70, which are made of tungsten, high-energy protons have much higher potential to generate secondary particles than the low-energy protons or even neutrons. For secondary particle yields, the C_V is dependent on the secondary particle species. For Z < 14, the C_V is less than 10%, but for heaver ions, some yields are less than 10 counts per 10^9 incidents, and this makes the C_V go up to 60%. For a few channels, the C_V is even 100% because only one count is observed.

EI evaluation of the secondary particles yield is shown in Table 2, and again 200 MeV protons and the LANSCE neutron have the most similar secondary particle yields. Similar to the case of LETs, 63 MeV proton has the worst EI in secondary particle yields. In Figure 6, the 63 MeV proton in the Z range between 18 and 25 contributed fewer secondary ions, which are copper spallation fragments. In Z between 65 and 70, the 63 MeV proton also has fewer secondary yields. Rummana et al. reported that secondary particle yields are higher for protons with higher energy [23].

Table 2. EI for secondary particle yields in layer structure without SiGe.

	LANSCE Neutron	63 MeV Proton	105 MeV Proton	150 MeV Proton	200 MeV Proton	230 MeV Proton
EI	0	0.560	0.544	0.543	0.405 *	0.443

* Indicates the best choice.

In Figure 6, 200 and 230 MeV protons typically gave more yields than the LANSCE neutron, but 63, 105, and 150 MeV protons gave less. It is possible that mixing different proton energies could lead to better neutron equivalence.

Therefore, from both Tables 1 and 2, 200 MeV proton radiation is closer to neutron radiation, and this concurs with suggestions from NASA (Washington, DC, USA).

3.3. LET Difference between Layer Structures with and without SiGe

The LET spectra for the LANSCE neutron and 63 and 230 MeV protons with and without SiGe were plotted in Figure 7. The difference in the overall spectra between the structures with and without SiGe is insignificant except for the LET value greater than 20 MeV-cm^2/mg. LANSCE neutron and 63 MeV proton give lower differential fluence when the SiGe layer is added, but 230 MeV proton gives a larger differential fluence. This is because with high-energy protons, the yields of alpha and light ions correlate with the Z of an incident target. Since Ge has a larger Z than most of the materials in our structures, being four times larger than the Z of O and 2.3 times larger than the Z of Si, we observed more counts for the structure with SiGe in the lower LET. With low-energy protons, however, it is difficult to have spallation and generate secondary ions [24]. In addition, Ge has a larger neutron absorption coefficient than silicon [25,26], hence some neutrons were absorbed by the SiGe layer. Therefore, in the case of the LANSCE neutron, the events with LETs less than 0.1 MeV-cm^2/mg are somewhat lower when SiGe is added.

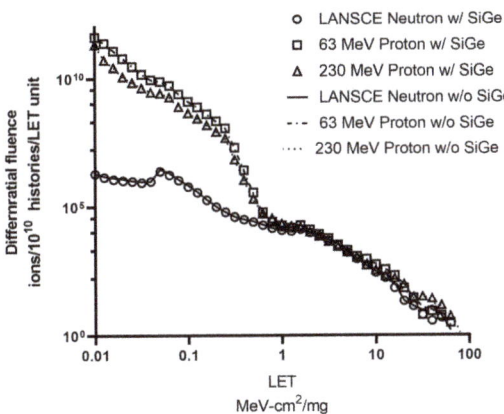

Figure 7. LET spectra of the structure with and without SiGe irradiated by 63 and 230 MeV protons and LANSCE neutron. The plot is in log-log scale.

3.4. Secondary Particle Yields Difference between Layer Structure with and without SiGe

Figure 8 shows the difference in secondary yields between the structure with and without SiGe layer. At Z between 30 and 32, which mostly come from Ge (Z = 32), the cases with SiGe give much higher yields than the cases without SiGe. The reason is that in cases without Ge, these secondary particles with Z between 30 and 32 can only be generated in the W layer (Z = 74) with a very low probability, thus the yield is much lower compared to the number of secondary particles of another Z. In Z between 35 and 45, the cases without SiGe give higher yields because the Ge can stop the movement of the heavy secondary particles due to the high Z and density of Ge. In comparison to the particle yields with Z < 30, however, the difference in secondary yield cannot be recognized, since these particles only make up less than a thousandth of all secondary particles.

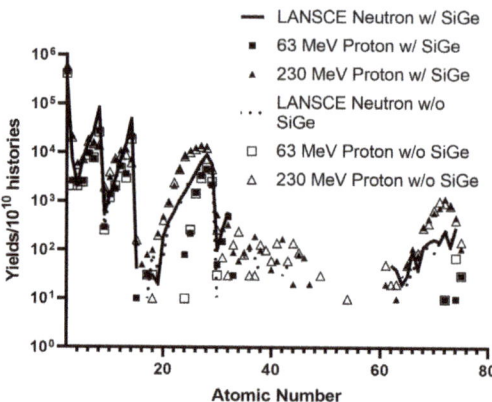

Figure 8. The secondary particle yields in the structure with and without SiGe irradiated by 63 and 230 MeV protons and LANSCE neutron.

Tables 3 and 4 giving the EI evaluation on LETs and secondary particle yields of the structure with SiGe layer irradiated by the LANSCE neutron and selected monoenergetic protons. Similar to the results shown in Tables 1 and 2, the 200 MeV proton gives the lowest EI which means the best neutron equivalence.

Table 3. EI for LET in layer structure with SiGe.

	LANSCE Neutron	63 MeV Proton	105 MeV Proton	150 MeV Proton	200 MeV Proton	230 MeV Proton
EI	0	0.237	0.256	0.258	0.228 *	0.257

* Indicates the best choice.

Table 4. EI for secondary particle yields in layer structure with SiGe.

	LANSCE Neutron	63 MeV Proton	105 MeV Proton	150 MeV Proton	200 MeV Proton	230 MeV Proton
EI	0	0.550	0.541	0.524	0.381 *	0.533

* Indicates the best choice.

3.5. Energy Deposited Difference between Neutrons and Protons

In addition to LET, the energy deposition in the total devices is important for the study of SEE. Since a proton is an ionizing radiation, it releases energy when it passes the target. The energy deposited in the detection layer was calculated and shown in Tables 5 and 6 for structures with and without SiGe.

Table 5. Energy deposition analysis results for the layer structure without SiGe for 10^{10} neutron/proton incident.

	LANSCE Neutron	63 MeV Proton	105 MeV Proton	150 MeV Proton	200 MeV Proton	230 MeV Proton
Energy deposited (GeV) $C_v <0.01\%$	2.7019	15,910	10,680	8213.4	6750.3	6165.4
LET >1 (counts) $C_v <0.7\%$	28,570	38,260	30,070	25,730	31,500	29,960
LET >10 (counts) $C_v <3\%$	1200	2500	1680	1370	2310	1920
Energy deposited/LET >1 (keV) $C_v <3\%$	94.6	41,500	35,500	31,900	21,400	20,500
Energy deposited/LET >10 (MeV) $C_v <4.2\%$	2.25	636	635	599	292	321

LET is in unit of MeV-cm^2/mg.

Table 6. Energy deposition analysis results for the layer structure with SiGe for 10^{10} neutron/proton incident.

	LANSCE Neutron	63 MeV Proton	105 MeV Proton	150 MeV Proton	200 MeV Proton	230 MeV Proton
Energy deposited (GeV) $C_v <0.01\%$	2.4761	15,956	10,706	8232.0	6765.3	6181.3
LET >1 (counts) $C_v <0.7\%$	27,460	36,700	30,350	26,010	32,040	31,510
LET >10 (counts) $C_v <3.2\%$	980	2360	1680	1280	2160	2360
Energy deposited/LET >1 (keV) $C_v <3.3\%$	90.2	43,500	35,300	31,600	21,100	19,600
Energy deposited/LET >10 (MeV) $C_v <4.6\%$	2.52	676	637	643	313	261

LET is in unit of MeV-cm^2/mg.

For the case without SiGe, Table 5 shows that 63 MeV protons are found to have more events with LET > 1 and 10 MeV-cm^2/mg than other incident particles, and they also cause the most energy deposition. This is expected as the lower energy will result in higher deposited energy since the stopping power is inversely proportional to the kinetic energy [24]. Calculating the energy deposited for generating an event, 200 MeV and 230 MeV give better efficiency (i.e., lower dose with higher LET counts). In comparison to neutrons, proton irradiation gives off over 200 times more energy deposited.

Compared to the results of structures with (Table 5) and without (Table 6) SiGe, both LANSCE neutron and protons show the differences on energy deposited and event number. For LET > 1 MeV-cm^2/mg, a decrease of 4% in the secondary particle yields is observed, and for LET > 0 MeV-cm^2/mg, the decrease is 20% in the LANSCE neutron case. These decreases are due to the presence of germanium which has a larger neutron absorption cross-section than other materials in this study [25,26].

In the proton cases, 63 MeV proton also has a 4% decrease in secondary yields from LET > 1 MeV-cm^2/mg and a 6% decrease for LET > 10 MeV-cm^2/mg. For 230 MeV proton, however, the secondary yield does not lead to a decrease, but rather to an increase both with LET > 1 MeV-cm^2/mg and LET > 10 MeV-cm^2/mg. This is because of the spallation which generates heavy secondary ions in the cases of proton irradiation. This spallation has its cross-section positively correlate to the incident energy and the Z of the target [19].

4. Conclusions

From the Monte Carlo studies in this work, in comparing the LET spectra and secondary particles yield from the proton and neutron radiation, we found that 200 MeV proton radiation has the closest resemblance to the neutron radiation, which concurs with suggestions from NASA. However, even with this close proton radiation, several differences present between proton and neutron radiation are as follows: First, proton radiation produces high secondary particles yield for LET > 15 MeV-cm^2/mg, and neutron hardly has LET > 10 MeV-cm^2/mg. Second, the secondary particles distribution is broader for the case of proton radiation. Third, proton radiation produces more secondary particles with Z > 14 whilst neutron radiation produces more secondary particles with 8 < Z < 14. Fourth, the energy deposited from proton radiation is around 300 times higher than neutron in the same flux. Fifth, the presence of SiGe does not affect the secondary particles yield for proton radiation, but it is decreased for neutron radiation. This implies that the strengthen of radiation robustness with the addition of SiGe cannot be seen with proton radiation.

In this study, we only focus on some commonly used monoenergetic protons in several testing protocols. As presented in the results, no monoenergetic proton can reproduce exactly the secondary particle LET and yield spectra to that of the LANSCE broad band neutron. However, it may be possible that mixing energies of protons can create better equivalence. Further study using range (energy) modulation technique to generate wider proton spectrum will be conducted to explore this possibility.

As LET and secondary particles yield can affect SEE testing, how the above-mentioned five differences will affect the SEE testing results are as of yet unknown. The answers can only be known through either subsequent proton or neutron testing or SEE simulation on semiconductor devices with LET and secondary particles yield distribution as obtained from the Geant4 simulation. On the other hand, as proton radiation seems to be more stringent than neutron radiation, one may use proton radiation tests as a higher calling to the radiation robustness of integrated circuits, with the expense of possibly higher design and fabrication costs. All these future works will be necessary in order to ascertain the equivalence or acceleration of neutron and protons radiation for SEE testing.

Author Contributions: Conceptualization, C.M.T. and T.-C.C.; methodology, C.-J.T. and C.-C.L.; Writing—original draft preparation, Y.C.; project administration, C.M.T.; writing—review and editing, C.M.T. and T.-C.C.; All authors have read and agreed to the published version of the manuscript.

Funding: This research was funded by Taiwan Semiconductor Manufacturing Company JDP project. And Chang Gung Medical Re-search Program under projects CIRPD1I0022, BMRP736, CIRPD2F0024, and CIRPD2I0012.

Acknowledgments: This work was technically supported by the Particle Physics and Beam Delivery Core Laboratory of the Institute for Radiological Research, Chang Gung University/Chang Gung Memorial Hospital.

Conflicts of Interest: All authors have no conflict of interest to the organizations mentioned in the paper.

References

1. Dong, A.X.; Gwinn, R.P.; Warner, N.M.; Caylor, L.M.; Doherty, M.J. Mitigating bit flips or single event upsets in epilepsy neurostimulators. *Epilepsy Behav. Case Rep.* **2016**, *5*, 72–74. [CrossRef] [PubMed]
2. Seifert, N.; Jahinuzzaman, S.; Velamala, J.; Ascazubi, R.; Patel, N.; Gill, B.; Basile, J.; Hicks, J. Soft error rate improvements in 14-nm technology featuring second-generation 3D tri-gate transistors. *IEEE Trans. Nucl. Sci.* **2015**, *62*, 2570–2577. [CrossRef]
3. Wie, B.S.; LaBel, K.A.; Turflinger, T.L.; Wert, J.L.; Foster, C.C.; Reed, R.A.; Kostic, A.D.; Moss, S.C.; Guertin, S.M.; Pankuch, M.; et al. Evaluation and Application of U.S. Medical Proton Facilities for Single Event Effects Test. *IEEE Trans. Nucl. Sci.* **2015**, *62*, 2490–2497. [CrossRef]
4. O'Neill, P.M.; Badhwar, G.D.; Culpepper, W.X. Internuclear cascade-evaporation model for LET spectra of 200 MeV protons used for parts testing. *IEEE Trans. Nucl. Sci.* **1998**, *45*, 2467–2474. [CrossRef] [PubMed]
5. Javanainen, A.; Malkiewicz, T.; Perkowski, J.; Trzaska, W.H.; Virtanen, A.; Berger, G.; Hajdas, W.; Lyapin, V.; Kettunen, H.; Mutterer, M.; et al. Linear Energy Transfer of Heavy Ions in Silicon. *IEEE Trans. Nucl. Sci.* **2007**, *54*, 1158–1162. [CrossRef]

6. Dodd, P.E.; Shaneyfelt, M.R.; Felix, J.A.; Schwank, J.R. Production and propagation of single-event transients in high-speed digital logic ICs. *IEEE Trans. Nucl. Sci.* **2004**, *51*, 3278–3284. [CrossRef]
7. Bagatin, M.; Gerardin, S.; Paccagnella, A.; Visconti, A.; Virtanen, A.; Kettunen, H.; Costantino, A.; Ferlet-Cavrois, V.; Zadeh, A. Single Event Upsets Induced by Direct Ionization from Low-Energy Protons in Floating Gate Cells. *IEEE Trans. Nucl. Sci.* **2017**, *64*, 464–470. [CrossRef]
8. Acosta Urdaneta, G.C.; Bisello, D.; Esposito, J.; Mastinu, P.; Prete, G.; Silvestrin, L.; Wyss, J. ANEM: A rotating composite target to produce an atmospheric-like neutron beam at the LNL SPES facility. In *International Journal of Modern Physics: Conference Series*; World Scientific Publishing Company: Singapore, 2016; Volume 44, p. 1660207. [CrossRef]
9. Agostinelli, S.; Allison, J.; Amako, K.A.; Apostolakis, J.; Araujo, H.; Arce, P.; Asai, M.; Axen, D.; Banerjee, S.; Behner, F.; et al. Geant4—A simulation toolkit. *Nucl. Instrum. Methods Phys. Res. Sect. A Accel. Spectrometers Detect. Assoc. Equip.* **2003**, *506*, 250–303. [CrossRef]
10. Foley, K.; Seifert, N.; Velamala, J.B.; Bennett, W.G.; Gupta, S. IRT: A modeling system for single event upset analysis that captures charge sharing effects. In Proceedings of the 2014 IEEE International Reliability Physics Symposium, Waikoloa, HI, USA, 1–5 June 2014; p. 5F.1. [CrossRef]
11. Weller, R.A.; Mendenhall, M.H.; Reed, R.A.; Schrimpf, R.D.; Warren, K.M.; Sierawski, B.D.; Massengill, L.W. Monte Carlo Simulation of Single Event Effects. *IEEE Trans. Nucl. Sci.* **2010**, *57*, 1726–1746. [CrossRef]
12. Allison, J.; Amako, K.; Apostolakis, J.; Arce, P.; Asai, M.; Aso, T.; Bagli, E.; Bagulya, A.; Banerjee, S.; Beck, B.R.; et al. Recent developments in Geant4. *Nucl. Instrum. Methods Phys. Res. Sect. A Accel. Spectrometers Detect. Assoc. Equip.* **2016**, *835*, 186–225. [CrossRef]
13. Truscott, P.; Lei, F.; Dyer, C.S.; Frydland, A.; Clucas, S.; Trousse, B.; Hunter, K.; Comber, C.; Chugg, A.; Moutrie, M. Assessment of neutron- and proton-induced nuclear interaction and ionization models in Geant4 for Simulating single event effects. *IEEE Trans. Nucl. Sci.* **2004**, *51*, 3369–3374. [CrossRef]
14. Geant4-Collaboration. *Book For Application Developers*, Rev10.4 ed.; CERN: Geneva, Switzerland, 2019.
15. Apostolakis, J.; Folger, G.; Grichine, V.; Heikkinen, A.; Howard, A.; Ivanchenko, V.; Kaitaniemi, P.; Koi, T.; Kosov, M.; Ribon, A.; et al. Progress in hadronic physics modelling in Geant4. *J. Phys. Conf. Ser.* **2009**, *160*. [CrossRef]
16. Tohka, J.; Reilhac, A. A Monte Carlo Study of Deconvolution Algorithms for Partial Volume Correction in Quantitative PET. In *2006 IEEE Nuclear Science Symposium Conference Record*; IEEE: Piscataway, HJ, USA, 2006; pp. 3339–3345.
17. Chiang, Y.; Tan, C.M.; Tung, C.-J.; Chao, T.-C. Lineal energy of proton in silicon by a microdosimetry simulation. *Radiat. Phys. Chem.* **2020**. submitted.
18. Hsing, C.-H.; Cho, I.C.; Chao, T.-C.; Hong, J.-H.; Tung, C.-J. GNP enhanced responses in microdosimetric spectra for 192Ir source. *Radiat. Meas.* **2018**, *118*, 67–71. [CrossRef]
19. Schwank, J.R.; Dodd, P.E.; Shaneyfelt, M.R.; Felix, J.A.; Hash, G.L.; Ferlet-Cavrois, V.; Paillet, P.; Baggio, J.; Tangyunyong, P.; Blackmore, E. Issues for single-event proton testing of SRAMs. *IEEE Trans. Nucl. Sci.* **2004**, *51*, 3692–3700. [CrossRef]
20. Erhardt, L.S.; Haslip, D.S.; Cousins, T.; Buhr, R.; Estan, D. Gamma enhancement of proton-induced SEE cross section in a CMOS SRAM. *IEEE Trans. Nucl. Sci.* **2002**, *49*, 2984–2989. [CrossRef]
21. Hiemstra, D.M.; Blackmore, E.W. Let spectra of proton energy levels from 50 to 500 mev and their effectiveness for single event effects characterization of microelectronics. *IEEE Trans. Nucl. Sci.* **2003**, *50*, 2245–2250. [CrossRef]
22. Turflinger, T.L.; Clymer, D.A.; Mason, L.W.; Stone, S.; George, J.S.; Koga, R.; Beach, E.; Huntington, K. Proton on Metal Fission Environments in an IC Package: An RHA Evaluation Method. *IEEE Trans. Nucl. Sci.* **2017**, *64*, 309–316. [CrossRef]
23. Rummana, A.; Barlow, R. Simulation and parameterisation of spallation neutron distributions. In *4th Workshop on ADS and Thorium*; SISSA Medialab: Trieste, Italy, 2017; p. 023.
24. Segrè, E.; Staub, H.; Bethe, H.A.; Ashkin, J. *Experimental Nuclear Physics. Volume I Volume I.*; John Wiley & Sons: New York, NY, USA; Chapman & Hall (in English): London, UK, 1953.

25. Hodgson, M.; Lohstroh, A.; Sellin, P.; Thomas, D. Neutron detection performance of silicon carbide and diamond detectors with incomplete charge collection properties. *Nucl. Instrum. Methods Phys. Res. Sect. A Accel. Spectrometers Detect. Assoc. Equip.* **2017**, *847*, 1–9. [CrossRef]
26. Aguayo, E.; Kouzes, R.; Orrell, J.; Reid, D.; Fast, J. *Optimization of the Transport Shield for Neutrinoless Double Beta-decay Enriched Germanium*; Pacific Northwest National Laboratory: Richland, WA, USA, 2012. [CrossRef]

© 2020 by the authors. Licensee MDPI, Basel, Switzerland. This article is an open access article distributed under the terms and conditions of the Creative Commons Attribution (CC BY) license (http://creativecommons.org/licenses/by/4.0/).

Article

Lineal Energy of Proton in Silicon by a Microdosimetry Simulation

Yueh Chiang [1], Cher Ming Tan [2,3,4], Chuan-Jong Tung [1,5], Chung-Chi Lee [1,5] and Tsi-Chian Chao [1,5,*]

1. Department of Medical Imaging and Radiological Sciences, College of Medicine, Chang Gung University, Kwei-Shan, Tao-Yuan 333, Taiwan; D0503203@cgu.edu.tw (Y.C.); cjtung@mail.cgu.edu.tw (C.-J.T.); cclee@mail.cgu.edu.tw (C.-C.L.)
2. Center for Reliability Sciences and Technologies, Chang Gung University, Kwei-Shan, Tao-Yuan 333, Taiwan; cmtan@cgu.edu.tw
3. Center for Reliability Engineering, Mingchi University of Technology, New Taipei City 243, Taiwan
4. Department of Urology, Chang Gung Memorial Hospital, Linkou 333, Taiwan
5. Particle Physics and Beam Delivery Core Laboratory, Institute for Radiological Research, Chang Gung University/Chang Gung Memorial Hospital, Linkou, Kwei-Shan, Tao-Yuan 333, Taiwan
* Correspondence: chaot@mail.cgu.edu.tw

Abstract: Single event upset, or Single Event Effect (SEE) is increasingly important as semiconductor devices are entering into nano-meter scale. The Linear Energy Transfer (LET) concept is commonly used to estimate the rate of SEE. The SEE, however, should be related to energy deposition of each stochastic event, but not LET which is a non-stochastic quantity. Instead, microdosimetry, which uses a lineal calculation of energy lost per step for each specific track, should be used to replace LET to predict microelectronic failure from SEEs. Monte Carlo simulation is used for the demonstration, and there are several parameters needed to optimise for SEE simulation, such as the target size, physical models and scoring techniques. We also show the thickness of the sensitive volume, which also correspond to the size of a device, will change the spectra of lineal energy. With a more comprehensive Monte Carlo simulation performed in this work, we also show and explain the differences in our results and the reported results such as those from Hiemstra et al. which are commonly used in semiconductor industry for the prediction of SEE in devices.

Keywords: single event effect; Monte Carlo simulation; microdosimetry; linear energy transfer; lineal energy

1. Introduction

Radiations exist around us in the air, and it includes photon, electron, neutron and alpha particle. These radiations are critical traditionally in the aerospace applications because they could render function loss for the satellites and even lead to material degradation [1], and they can be ignored at the sea level. However, as microelectronic devices are scaling down aggressively with the advancement in semiconductor technology, nano-meter devices are now susceptible to these radiations, and they can no longer be overlooked. These radiations can interact with semiconductor devices and cause damages to the devices or functional errors to their associated circuits [2].

The effect of these radiations on microelectronic devices can be categorized into two different effects, namely a cumulative effect termed as Total Ionization Dose (TID) effect, and a stochastic effect, termed as Single Event Effect (SEE) [3]. SEE describes the event of electronic device malfunction or failure caused by "single" radiation hits. SEE is critical for devices where their rebooting is difficult or even impossible, and examples of such devices are pacemakers, autopilot systems, and microelectronics used in space missions. Lineal Energy Transfer (LET) is widely used as a key index for SEE prediction [4,5].

LET is a non-stochastic quantity which gave an expectation value of energy imparted from particle to a local site. International Commission on Radiation Units and Measurements (ICRU) gave the definition of LET of a particle (Equation (1)) in its report 16 [6] as the quotient of dE by dl, where dl is the distance traversed by this particle and dE is the mean energy-loss owing to collisions with energy transfers less than some upper limit Δ, which is the cut off energy of delta ray.

$$L_\Delta = \left(\frac{dE}{dl}\right)_\Delta \tag{1}$$

There is a long history of studies on LET in microelectronic applications. O'Neill et al. conducted a series of studies on LET with secondary ions generated by the irradiation of silicon with protons, and they showed that proton testing is suitable to screen microelectronic devices for low earth orbit susceptibility to heavy ions [7]. Hiemstra et al. provided the LET spectra of protons in 50–500 MeV, which are commonly used for SEE testing, and gave the scaled factor of proton fluence for Geosynchronous orbits by a simplified Monte Carlo simulation [8]. Biersack and Ziegler carried out a comprehensive study of the ion ranges and stopping powers in solids and provide a useful toolkit called Stopping and Range of Ions in Matter (SRIM). This toolkit allows the radiation particles to be modelled in a simple microelectronic structure [9,10].

Dodd et al. provided the trend of LET thresholds of SEE with feature sizes from 1 to 0.1 µm [11]. In their results, the threshold decreases when the feature size decreases due to the following. Firstly, when the sizes of the devices are smaller, the critical charges, i.e., the amount of charges to change the performance of the devices will be smaller. Secondly, with a smaller target, the energy deposition of each specific radiation particle will have a greater deviation, and there is a chance of a huge energy deposition leading to a huge induced charge.

Single Event Rate (SER) can be directly tested under specific protocols [12] or modelled with known natural radiation environment for these cosmic rays [13]. JEDEC Solid State Technology Association give the protocol JESD57A [12] to guide the procedures for the SEE tests. These tests are usually expensive and difficult. Another way to predict single event rate is to convolute the LET spectra in target orbital with the single event cross sections (either for a device or averaged per bit) from either experiment or simulation. In this method, single event cross sections versus LET is usually a single value function. Warren used a set of heavy ion with specific kinetic energy to derive the single event cross sections and predict the single event upset rate of a 0.25 µm CMOS SRAM well [14].

As the feature size reduced to a few nanometres, this "single-value assumption" may no longer valid. Warren et al. found that in a 90 nm CMOS irradiated with heavy ions, although the effective LET is the same, the single event cross sections can have more than three order of magnitude differences [15]. In other words, LET distributions can be different in various feature sizes of the devices under an exact irradiation condition. With today nano-meter microelectronic devices, the variability of LET from a given radiation can be very large, and this leads to the necessity to use the concept of lineal energy distribution to predict SEE in a given device.

In fact, LET is commonly used in the cm/mm scale for deterministic events. For microelectronics, microdosimetry is more appropriate to describe the stochastic events. Down to nanometre scale, many researchers used realistic track structure to predict local effect of different radiations [16–19]. This track structure approach will be practical in the simulation, but it is still difficult to compare with measurements. In addition, these track structure codes can only simulate interactions in limited materials and semiconductor materials is yet to be included. Furthermore, the track structure will only be meaningful when detailed geometry of a specific microelectronic device is given. On the other hand, microdosimetry is easier to be modelled for radiation effect prediction [20]. We have used this microdosimetry approach to investigate the equivalence of neutrons and protons in SEEs testing [21].

ICRU report 16 considered that the local energy density and individual event size will be more relevant than LET for microscopic structures [6]. These concepts of local energy density and individual event size have been clarified in the ICRU report 36 for the introduction of microdosimetry [22]. The basic concepts of microdosimetry are developed by Rossi and co-workers [23,24], and they used two stochastic quantities, specific energy (z) and lineal energy (y) to replace dose and LET. Lineal energy (y) of microdosimetry is a way to estimate radiation quality/impact microscopically [22], and it is defined as,

$$y = \frac{\varepsilon}{\bar{l}} = \frac{\sum_i \epsilon_i}{\bar{l}} \qquad (2)$$

where the parameters in the equation is defined below.

When a radiation particle enters a target, it will interact with the target material along its track in the material. Energy deposit ϵ_i is defined as the energy deposited in i^{th} "Single" interaction. The sum of the ϵ_i in the target within one track is defined as the energy imparted (ε). Mean chord length ($\bar{l}$) is the mean length of randomly oriented chords in a given volume [22].

In large size target, the number of interactions will be large, and the distribution of y will converge to an expected value, which is LET. However, at the nanometric scale, the number of interactions is small, and only very few interaction events occur in a radiation track. Thus, y would have a wide distribution. In this case, providing discrete values of LET will be meaningless because different y values would have different contributions to SEEs and the relationship is non-linear.

Monte Carlo simulation is commonly used in microdosimetry studies [25,26], and it relies on many physical models among which nuclear interaction physics is the most important for microdosimetry study on SEE. However, a Monte Carlo simulation with detailed nuclear interaction is very time and computation resources consuming. When computing power was expensive in the past, secondary particles generation and their transport in silicon could not be included in the simulation. For example, Hiemstra et al. tried to use a simplified Monte Carlo method, which allow only one filial of secondary particles to compute their corresponding ranges. They then used the range to calculate LET and predict SEE [8]. However, the secondary particles generated under proton irradiation are usually quite unstable and will decay or have further interactions with silicon, which may generate light ions with lower energy and higher stopping powers. All these ions can affect the SEE. An example of such can be found by Ying et al. who found that carbon ions can fragment into several secondary particles during transportation [27].

Recently, there is an increasing number of Monte Carlo based SEE evaluation toolkits. Intel developed the Intel Radiation Tool (IRT) based on Geant4 [28]. Reed and colleagues in Vanderbilt University has also developed another Geant4 based Monte Carlo Radiative Energy Deposition (MRED) Code [29]. In Geant4 Space Users' Workshop and Nuclear & Space Radiation Effects Conference (NSREC), using Monte Carlo technique to predict radiation effect in semiconductor is a routine discussion. However, within the Monte Carlo framework, there are hundreds of parameters that could affect the results, and there is no conclusion on which setting is more accurate. Besides, the setting should be different for different purpose.

In this work, a Geant4 [16] based Monte Carlo simulation is performed to simulate the effect of proton beams on silicon. Different sensitive volume thicknesses are compared. The lineal energy is also analysed in detail for each particle crossing the sensitivity volume, and the physical models for intra-nuclear interaction in Geant4 are also compared.

2. Simulation Setup

The simulation in this study is based on Geant4 10.05 [16]. The physics model of electromagnetic processes in this simulation is aligned with G4EmStandardPhysics_option4. For electrons, due to their low LET, they cannot contribute to high y effects [30], and thus the step function (parameter for step size R) can be set such that R over range = 0.01 and final range = 1 nm to speed up the simulation. For hadrons that have high LET and they are therefore more important in this study, the R over range = 0.00001 and the final range = 0.1 nm are set to improve the accuracy while keeping the efficiency of the Monte Carlo simulation.

For the intranuclear cascade, the Bertini's model is widely used [31]. The nucleon spectra from continuum-state transitions for protons on complex nuclei are calculated using the intranuclear-cascade approach. However, the model is commissioned in high-energy physics. An alternative method is a detailed three-dimensional model of the nucleus, and is based exclusively on the binary scattering between reaction participants and nucleons within this nuclear model, called binary cascade model [32]. Wright et al. showed that, at a few hundreds of MeV, the binary cascade model can predict the secondary particles yields comparable to experimental data effectively [33,34]. Additionally, there are several evaluated and experimental nuclear cross-section data banks, such as TENDL [35], EXFOR [36] and JEFF [37] which are precise but they are only for limited materials.

The intra-nuclear model, which controls secondary yields, will be compared in Geant4 using two different models. The Bertini cascade model which is widely used in SEEs studies, and the binary cascade which is found to be better for incident particles of several hundreds of MeV [34]. In addition, this work also considers the impact of high precision (HP) model using the JEFF 3.3 extension cross-section table.

The geometry in this project is a Multi-layer cylinder as shown in Figure 1. The diameter of the cylinder is 1 mm in order to ensure that most secondaries will not go out of the boundary. A 30-µm of pure silicon layer is used to generate secondary particles. Under the silicon layer is the sensitive volume (SV). The SV thickness was used to calculated lineal energy instead of mean chord length in the study. As described previously, the mean chord length concept is used better for randomly oriented chords in a given volume. The primary particles simulated in this study do not have isotropic distribution, and they are mostly incident perpendicular to the SV. Although ICRU report 36 [22] recommended using 4 V/A to calculate mean chord length, Horowitz and Dubi [38] found that real average path length was smaller than mean chord length for non-isotropic irradiation. If we calculated mean chord length according to the 4 V/A approach, the mean chord length of our SV with 1 mm diameter and 100 nm thickness will be 200 nm. However, the average track length in this thickness from the simulation is 104.9 nm, which is less than 5% difference to the SV thickness. Hence we use the SV thickness instead of the mean chord length to calculate y.

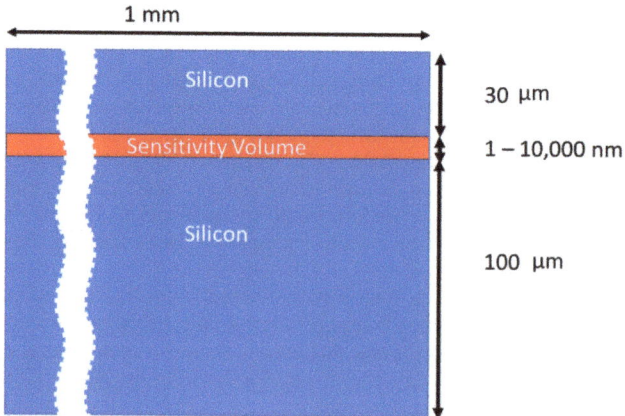

Figure 1. The geometry setup in this study. The silicon is with natural isotope abudence, density is 2330 mg/cm^3 and mean extiation potetial I = 173 eV.

The species and kinetic energy of each particle entering the SV are recorded as secondaries. The energies imparted from all the tracks crossing through the SV are summed up and divided by the thickness of SV to calculate the lineal energy (y). To compare the effect of SV thickness, six thicknesses, namely 1, 10, 30, 100, 1000 and 10,000 nm are simulated.

In addition, the lineal energy and secondary particle yields from the thickness of 100 nm results are further used to compared with the results from Hiemstra et al. for compatibility investigation.

For the SEEs prediction purposed, the unit of y is presented in MeV-cm^2/mg instead of keV/μm, which is the traditional unit in microdosimetry. In silicon, 1 MeV-cm^2/mg is equal to 233 keV/m. In conventional microdosimetry, the scale usually covers from 10^{-3} to 10^3 keV/μm. To present the entire y spectrum, semi-log scale is usually used. However, for SEEs, the LET usually is in the range of 10^{-1} to 10^2 MeV-cm^2/mg, and differential fluence is used instead of probability density function of y.

3. Results and Discussions

3.1. Effect of SV Thickness on y Distribution

Dodd et al. [11] showed that smaller feature sizes have smaller SEE thresholds. In this study, six SV thicknesses were simulated to mimic different feature sizes. Figure 2 shows the effect of different SV thickness on the lineal energy (y) distribution. In the 10,000 nm case (green line), no secondary particle has y > 10 MeV-cm^2/mg. The ε for most secondary particles (more than 99.999%) were less than 2.33 MeV, corresponding to 1 MeV-cm^2/mg.

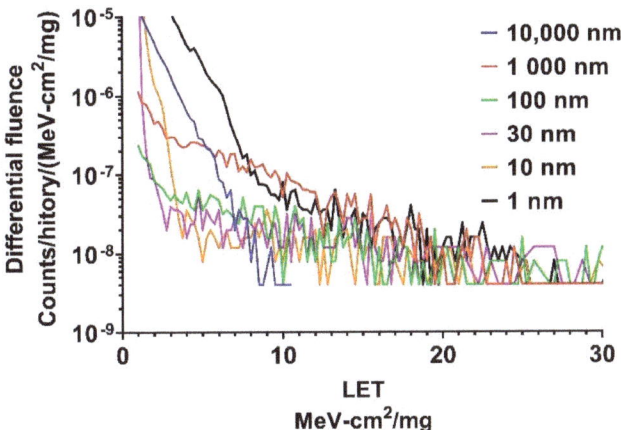

Figure 2. Lineal energy spectra of different sensitive volumes in silicon irradiated by a 200 MeV proton beam.

Let us consider the secondaries with y > 10 MeV-cm^2/mg for other SV thickness. Their differential fluences decreases as the SV thickness decreases from 1000 nm to 10 nm, and this trend can be explained as follows. The y value of a particle is usually maximized when the SV thickness is close to the particle range, owing to a pronounced nature named Bragg peak. The Bragg peak describes that the heavy charged particles have little energy loss as they enter the material and then peak before the end of their path. In Figure 3, the cumulative distribution function of kinetic energy of secondary particles generated by 200 MeV proton irradiated on silicon is plotted. In this figure, more than 80% of secondary particles with atomic number (Z) larger than 2 have the kinetic energy lower than 10 MeV and 99% lower than 20 MeV. The continuous slowing down approximation (CSDA) ranges of these particles are about few thousand nm. During the slowing down process, the stopping power of particles increases as they loss their kinetic energy. The stopping power reach maximum at the end of the particle's range. When the SV size close to the particle range, it can cover the maximum stopping power. Since the particle range is approximately 1 μm, the y values for the case of SV thickness = 1000 nm is larger than that for SV = 100 nm, and then 10 nm. On the other hand, when the thickness decreases to 1 nm, the differential fluences increases significantly, and the reason will be discussed later using a CROSSER and STOPPER theory [18]. As the boundary of the CROSSER and STOPPER will be visible only at lower LET, hence this theory does not apply in this discussion where $y > 10$ MeV-cm^2/mg.

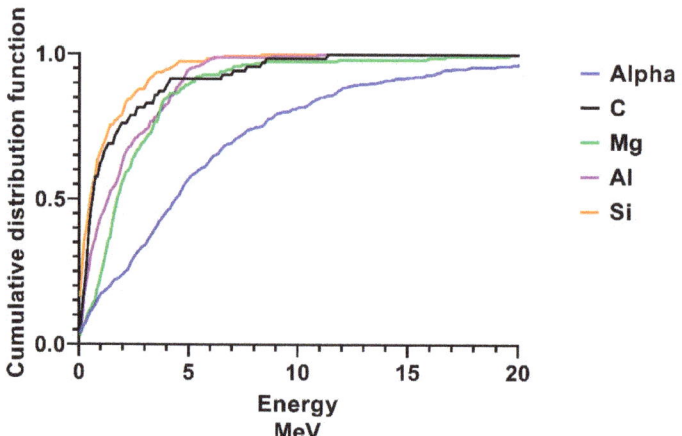

Figure 3. Cumulative distribution function of kinetic energy of secondary particles generated by 200 protons irradiated on silicon.

For the case of $y < 10$ MeV-cm^2/mg, Figure 2 shows that the differential fluences decrease first when thickness decreased from 10,000 nm to 1000 nm, and then 100 nm. However, it increases when SV thickness continues to decrease from 100 nm to 10 nm and 1 nm. To understand the above-mentioned trends, additional SV thicknesses of 20, 40 and 50 nm are simulated respectively. Figure 4 plots the spectra in log scale, which can present the events with smaller y. In these spectra, "shoulders" can be seen for SV thickness from 10–100 nm, and the "shoulder" left shift with the SV thickness. The shoulder of 10 nm case can be found in Figure 2 which is around y = 2–3 MeV-cm^2/mg. The presence of this "shoulder" can be understood from the CROSSER and STOPPER theory [18] as elaborated below.

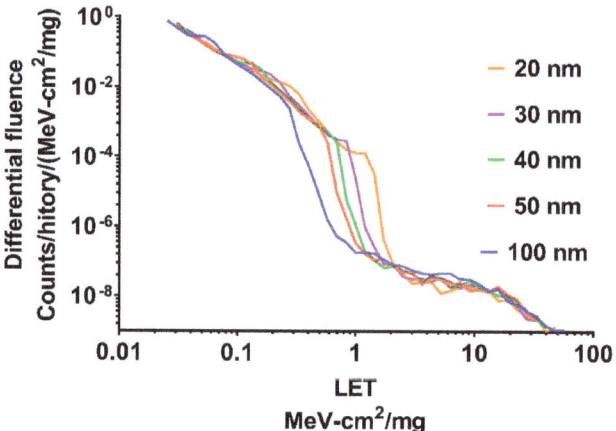

Figure 4. Lineal energy spectra of different sensitive volumes in silicon irradiated by a 200 MeV proton beam (log y scale).

When particles travel through different SVs, they can end up into either CROSSERs or STOPPERs. A CROSSER refers to particle with high enough energy that can go through the entire SV, and a STOPPER refers to particle that stop inside SV because of its insufficient energy [18]. In Figures 2 and 4, the events on the left side of shoulder are contributed from both the CROSSER and STOOPER, but the right side can only be from the CROSSER. Since the definition of lineal energy is the energy imparted in the SV divided by the mean chord length, if the particle stopped inside the cavity, then y will be underestimated. In thicker SV, because the energy required to go through the SV is larger and more particle may become STOPPER, its y will be underestimated significantly. In Figures 2 and 4, the y-distribution of 1 nm is fairly smooth and no shoulder is observed, which means that most of the particles are CROSSER and the observed y are more realistically estimated. In SV thicker than 100 nm, the shoulder is unclear because most of the particles are STOPPERs.

Similarly, for the case of 10,000 nm, a y cut off is found and the y of heavy secondary particles is underestimated because it cannot go through the SV. In other words, the CROSSER or STOPPER theory also answers the question on the increasing differential fluence in the 1 nm case. With the shift of the shoulder to the right with decreasing SV thickness, the STOPPER becomes CROSSER, especially in the 1 nm case, and almost all particles are CROSSER.

3.2. Lineal Energy Contribution from Various Secondary Species

Since Single Event Rate (SER) is a function of LET [39], or y in this study, it is important to understand the y contribution from various secondary species for radiation hardening design. Figure 3 shows the y spectra of various types of secondary particles in pure silicon irradiated with 200 MeV protons.

The y spectra in silicon irradiated by a 200 MeV proton beam in this study was quite different from the results from Schwank [40] and Hiemstra [8]. In Schwank [40] and Hiemstra [8], their LET has a cut off around 15–16 MeV-cm^2/mg. However, there are few events with y as high as 60 MeV-cm^2/mg as observed in our study. Furthermore, the spectra in their studies have a plateau before 10 MeV-cm^2/mg and drop rapidly after 10 MeV-cm^2/mg. In our case, the counts dropped quickly before 2 MeV-cm^2/mg, and it became a straight line from 2 to 25 MeV-cm^2/mg, followed by a long tail after 25 MeV-cm^2/mg. To better understand the above-mentioned discrepancies, the y spectrum was separated by different contributors with various Z values.

For y < 2 MeV-cm^2/mg, the major contributors are particles with Z = 1 and 2. As they are light ions, their contributions to LET spectra dropped sharply and disappeared for y > 10 MeV-cm^2/mg.

For y between 2 and 10 MeV-cm^2/mg, the contribution is mixed with low Z and high Z secondaries. In this region, contributions from low Z secondaries decrease sharply and high Z secondary particles became major contributors after y > 5 MeV-cm^2/mg. These high Z secondaries were from heavy ions generated by (p, x) reactions. The y distribution of high Z secondaries had a peak around y = 5 MeV-cm^2/mg with few counts in low y and a long tail in the high y region, extending to 50 to 60 MeV-cm^2/mg.

Therefore, the overall y spectra of 200 MeV proton irradiated on silicon can be separated into three parts as follows. First part is y < 2.5 MeV-cm^2/mg contributed from light ions; second part is 5 MeV-cm^2/mg < y < 10 MeV-cm^2/mg contributed by a mixture of both light and heavy ions and third part is y > 10 MeV-cm^2/mg contributed only by heavy ions. In contrast to the works of Schwank et al. and Hiemstra et al. [8,40], the LET spectra of their studies have a cut off around 15–16 MeV-cm^2/mg but not in our study.

This difference of our work and their work is because Schwank and Hiemstra used a reference Table to calculate the LET from energy fluence instead of calculating the energy deposition, and such method is only applicable for large target where the number of interactions is large enough that y can convergence to a single value which is LET. Unfortunately, in tiny SV, the y distribution is quite board when the SV is thin and using single expected value of LET becomes meaningless. Moreover, the reference Table used in their study has the cut off around 15–16 MeV-cm^2/mg.

The differential fluences for LET < 5 MeV-cm^2/mg are significantly different between our results and those of Hiemstra et al. [8]. In Figure 7 of Hiemstra's work. the differential fluence is around 3000/10^{10} incident protons at LET = 1 MeV-cm^2/mg and 1000/10^{10} at 5 MeV-cm^2/mg. However, in our study, the differential fluence is 2400/10^{10} at 1 MeV-cm^2/mg and 360/10^{10} at 5 MeV-cm^2/mg as shown in the Figure 4. The slope in our study is about twice that of the Hiemstra's result because Hiemstra et al. did not consider the light secondaries such as protons and helium. In our study, the y contribution from each specific secondary particle is separate recorded, and they are plotted in Figure 5, where the proton and helium contribute more than 99% of the events in the region of y < 5 MeV-cm^2/mg.

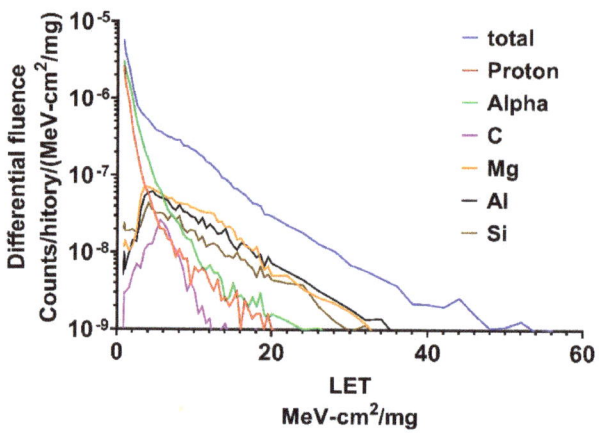

Figure 5. y spectra in 100 nm silicon irradiated by a 200 MeV proton beam.

In previous SEEs studies, LET was usually treated as a single value for each incident particle with specific kinetic energy. However, Schrimpf et al. [4] and Shaneyfelt et al. [5] presented that, in some cases, LET verses SEEs cross-section is not a one to one function. In other words, the SEEs cross-section sometimes has large variation even though the LET is the same. In Table 1, the calculated LET of each type of secondary particle via the mean of its kinetic energy are listed. The calculation is done by two commonly used tools, namely LET124 and SRIM. LET124 is from the Tandem Van de Graaff Accelerator Facility at Brookhaven National Laboratory's and SRIM is from Ziegler et al. [7].

Table 1. The calculated LET using mean energy of secondary particles generated by 200 MeV proton irradiate on silicon.

Z #	Mean Energy (MeV)	LET (LET124) * (MeV-cm^2/mg)	LET (SRIM-2013) ** (MeV-cm^2/mg)
2	5.32	0.5949	0.588
3	2.83	1.878	2.134
4	3.00	3.003	3.138
5	1.93	4.292	4.197
6	1.53	5.118	4.853
7	2.84	6.41	6.006
8	3.66	7.416	7.126
9	3.85	8.381	8.162
10	4.60	9.358	8.172
11	3.55	9.69	8.544
12	2.76	9.684	8.186
13	1.98	8.616	6.748
14	1.20	6.719	5.973

* Calculated by LET124 from Tandem Van de Graaff Accelerator Facility, Brookhaven National Laboratory, USA. ** Calculated by SRIM [9].

In the Table 1, even though both calculations are based on Bethe stopping formula but with different correction factor, the results can have more than 20% difference. When we look back to Figure 5, which plots the y in the SV for each kind of secondary particles, the y is not a single value or a symmetric distribution such as Gaussian distribution. Instead, the distribution is focus on low y region and has a very long tail in high y region. Take Z = 12 which is magnesium for example, the calculated LET is around 7 MeV-cm^2/mg, but the maximum y can be larger than 50 MeV-cm^2/mg, which is seven times higher than the mean. Therefore, using single value LET to estimate the SEEs is risky because the LET can only present the low y region, but the event which contribute SEEs can mostly from the high y tail. Due to the counts in high y region is quite less as compared to low y region, it has only little effect on the mean but may have significant effect on SEEs cross-section.

3.3. Effect of Various Physics Models on Secondary Yields

Raine and Jay et al. showed that the displacement damage in silicon from single particle interaction is dependent on the secondary particles type and energy deposition [41–43]. In this study, we also investigate if different physics models may affect the secondary particle yield. Figure 6 shows the secondary particle yields using various intra-nuclear cascade models in Geant4, and their comparison to Hiemstra's results [8]. The Hiemstra results come very close to the results of using the Bertini model. However, the peak in Z = 6 in this study was much higher than that from Hiemstra's result. In Hiemstra's study, all the secondary particles are from primary proton and they were calculated by the LAHET code system using Bertini cascade model. In the LET calculation part, Hiemstra's study did not consider the decay and further nuclear interaction. In our study, the primary proton will generate the first filial and first filial can still has a nuclear interaction and decay to generate the second or further filial. That is, a heavier secondary particle (Z = 12, 13, or 14) may break into two middle-sized fragments (Z = 6) during its transportation as also shown by McNulty's result [44].

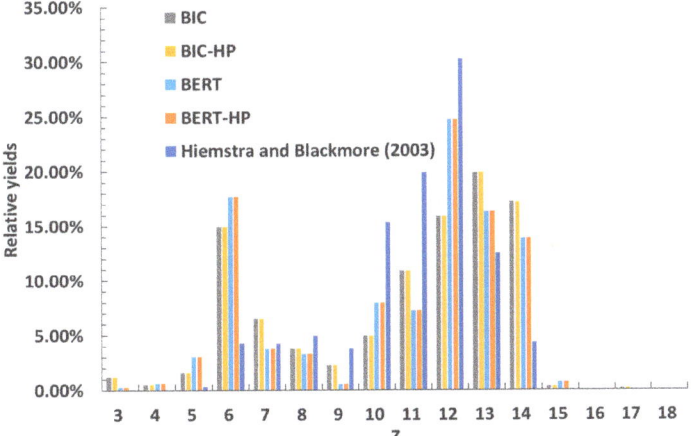

Figure 6. Secondary particle yields in 100 nm silicon irradiated by a 200 MeV proton beam using various physics models. BIC represents the Binary cascade model. BERT represents Bertini cascade model. HP represents high precision add-on.

Compared to the differences between the binary and the Bertini cascade, it is difficult to say that they are systematically different in Figure 6 and measurements are needed. In addition, for 10^9 incident particles, the high-precision physical model in Geant4 gave a negligible difference in secondary particle yields.

4. Conclusions

The concept of LET has been used to evaluate the Single Event Effect of radiation on semiconductor devices. In this work, we demonstrated that the conventional use of single value of LET is no longer applicable when the device dimension is scaling down to nano-material range. Instead, microdosimetry concept where the distribution of lineal energy will be necessary. We also performed a detailed microdosimetry Monte Carlo simulation of secondary particles from protons irradiating on silicon, and the results are shown to be very different from the previously reported work by Hiemstra and others with detail explanation. In particular, as compared to the simplified Monte Carlo results from Hiemstra [8], our results give much more low Z secondary particles which may not affect larger feature size but it will affect feature in nano-meter scale [45]. However, this study only focuses on a simplified hypothetical geometry. For realistic geometry, track structure technique should be used to estimate the geometrical distribution of energy deposition. Experimental verification of this difference will be our future work.

Author Contributions: Conceptualization, C.M.T. and T.-C.C.; methodology, C.-J.T. and C.-C.L.; Writing—original draft preparation, Y.C.; project administration, C.M.T.; writing—review and editing, C.M.T. and T.-C.C. All authors have read and agreed to the published version of the manuscript.

Funding: This research was funded by Taiwan Semiconductor Manufacturing Company JDP project. And Chang Gung Medical Research Program under projects CIRPD1I0021, BMRP736, CIRPD2F0024, and CIRPD2I0012.

Acknowledgments: This work was technically supported by the Particle Physics and Beam Delivery Core Laboratory of the Institute for Radiological Research, Chang Gung University/Chang Gung Memorial Hospital.

Conflicts of Interest: All authors have no conflict of interest to the organizations mentioned in the paper.

References

1. Garoli, D.; De Marcos, L.V.R.; Larruquert, J.I.; Corso, A.J.; Zaccaria, R.P.; Pelizzo, M.G. Mirrors for Space Telescopes: Degradation Issues. *Appl. Sci.* **2020**, *10*, 7538. [CrossRef]
2. Insoo, J. Effects of secondary particles on the total dose and the displacement damage in space proton environments. *IEEE Trans. Nucl. Sci.* **2001**, *48*, 162–175. [CrossRef]
3. Tan, F.; Huang, R.; An, X.; Wu, W.; Feng, H.; Huang, L.; Fan, J.; Zhang, X.; Wang, Y. Total ionizing dose (TID) effect and single event effect (SEE) in quasi-SOI nmOSFETs. *Semicond. Sci. Technol.* **2013**, *29*, 015010. [CrossRef]
4. Schrimpf, R.D.; Warren, K.M.; Weller, R.A.; Reed, R.A.; Massengill, L.W.; Alles, M.L.; Fleetwood, D.M.; Zhou, X.J.; Tsetseris, L.; Pantelides, S.T. Reliability and radiation effects in IC technologies. *IEEE Int. Reliab. Phys. Symp.* **2008**, 97–106. [CrossRef]
5. Shaneyfelt, M.R.; Schwank, J.R.; Dodd, P.E.; Felix, J.A. Total Ionizing Dose and Single Event Effects Hardness Assurance Qualification Issues for Microelectronics. *IEEE Trans. Nucl. Sci.* **2008**, *55*, 1926–1946. [CrossRef]
6. International Commission on Radiation Units and Measurements. *Linear Energy Transfer, in Report 16*; International Commission on Radiation Units and Measurements: Stockholm, Sweden, 2016; p. NP.
7. O'Neill, P.; Badhwar, G.; Culpepper, W. Risk assessment for heavy ions of parts tested with protons. *IEEE Trans. Nucl. Sci.* **1997**, *44*, 2311–2314. [CrossRef]
8. Hiemstra, D.; Blackmore, E. Let spectra of proton energy levels from 50 to 500 MeV and their effectiveness for single event effects characterization of microelectronics. *IEEE Trans. Nucl. Sci.* **2003**, *50*, 2245–2250. [CrossRef]
9. Ziegler, J.F.; Ziegler, M.; Biersack, J. SRIM—The stopping and range of ions in matter (2010). *Beam Interact. Mater. Atoms* **2010**, *268*, 1818–1823. [CrossRef]
10. Biersack, J.P.; Ziegler, J.F. The Stopping and Range of Ions in Solids. In *Ion Implantation Techniques*; Springer Nature: London, UK, 1982; pp. 122–156.
11. Dodd, P.; Shaneyfelt, M.; Felix, J.; Schwank, J. Production and propagation of single-event transients in high-speed digital logic ICs. *IEEE Trans. Nucl. Sci.* **2004**, *51*, 3278–3284. [CrossRef]
12. JEDEC Solid State Technology Association. *Test. Procedures for the Measurement of Single-Event Effects in Semiconductor Devices from Heavy Ion. Irradiation, in JESD57A*; JEDEC Solid State Technology Association: Arlington, VA, USA, 2003.
13. Pickel, J.C. Single-event effects rate prediction. *IEEE Trans. Nucl. Sci.* **1996**, *43*, 483–495. [CrossRef]
14. Warren, K.M.; Weller, R.A.; Sierawski, B.D.; Reed, R.A.; Mendenhall, M.H.; Schrimpf, R.D.; Massengill, L.W.; Porter, M.E.; Wilkinson, J.D.; Label, K.A.; et al. Application of RADSAFE to Model the Single Event Upset Response of a 0.25 μm CMOS SRAM. *IEEE Trans. Nucl. Sci.* **2007**, *54*, 898–903. [CrossRef]

15. Warren, K.M.; Sierawski, B.D.; Reed, R.A.; Weller, R.A.; Carmichael, C.; Lesea, A.; Mendenhall, M.H.; Dodd, P.E.; Schrimpf, R.D.; Massengill, L.W.; et al. Monte-Carlo Based On-Orbit Single Event Upset Rate Prediction for a Radiation Hardened by Design Latch. *IEEE Trans. Nucl. Sci.* **2007**, *54*, 2419–2425. [CrossRef]
16. Bernal, M.A.; Bordage, M.-C.; Brown, J.; Davidkova, M.; Delage, E.; El Bitar, Z.; Enger, S.; Francis, Z.; Guatelli, S.; Ivanchenko, V.N.; et al. Track structure modeling in liquid water: A review of the Geant4-DNA very low energy extension of the Geant4 Monte Carlo simulation toolkit. *Phys. Medica* **2015**, *31*, 861–874. [CrossRef] [PubMed]
17. Incerti, S.S.; Baldacchino, G.; A Bernal, M.; Capra, R.; Champion, C.; Francis, Z.; Guèye, P.; Mantero, A.; Mascialino, B.; Moretto, P.; et al. THE GEANT4-DNA PROJECT. *Int. J. Model. Simul. Sci. Comput.* **2010**, *1*, 157–178. [CrossRef]
18. Incerti, S.; Ivanchenko, A.; Karamitros, M.; Mantero, A.; Moretto, P.; Tran, H.N.; Mascialino, B.; Champion, C.; Ivanchenko, V.N.; Bernal, M.A.; et al. Comparison ofGEANT4very low energy cross section models with experimental data in water. *Med. Phys.* **2010**, *37*, 4692–4708. [CrossRef]
19. Incerti, S.; Kyriakou, I.; Bernal, M.A.; Bordage, M.-C.; Francis, Z.; Guatelli, S.; Ivanchenko, V.; Karamitros, M.; Lampe, N.; Lee, S.B.; et al. Geant4-DNA example applications for track structure simulations in liquid water: A report from the Geant4-DNA Project. *Med. Phys.* **2018**, *45*, e722–e739. [CrossRef]
20. Hawkins, R.B. A microdosimetric-kinetic theory of the dependence of the RBE for cell death on LET. *Med. Phys.* **1998**, *25*, 1157–1170. [CrossRef]
21. Chiang, Y.; Tan, C.M.; Chao, T.-C.; Lee, C.-C.; Tung, C.-J. Investigate the Equivalence of Neutrons and Protons in Single Event Effects Testing: A Geant4 Study. *Appl. Sci.* **2020**, *10*, 3234. [CrossRef]
22. Dennis, J.A. Book review Microdosimetry. ICRU Report No. 36. pp. 118, 1983. (International Commission on Radiation Units and Measurements, Bethesda, Md.) $18.00. *Br. J. Radiol.* **1985**, *58*, 250. [CrossRef]
23. Rossi, H.; Failla, G. Neutrons: Dosimetry. *Med. Phys.* **1950**, *2*, 603–607.
24. Rossi, H.H. Specification of Radiation Quality. *Radiat. Res.* **1959**, *10*, 522. [CrossRef]
25. Conte, V.; Colautti, P.; Grosswendt, B.; Moro, D.; De Nardo, L. Track structure of light ions: Experiments and simulations. *New J. Phys.* **2012**, *14*. [CrossRef]
26. Valentín, A.; Raine, M.; Sauvestre, J.-E.; Gaillardin, M.; Paillet, P. Geant4 physics processes for microdosimetry simulation: Very low energy electromagnetic models for electrons in silicon. *Beam Interact. Mater. Atoms* **2012**, *288*, 66–73. [CrossRef]
27. Ying, C.K.; Bolst, D.; Tran, L.T.; Guatelli, S.; Rosenfeld, A.B.; A Kamil, W. Contributions of secondary fragmentation by carbon ion beams in water phantom: Monte Carlo simulation. *J. Physics Conf. Ser.* **2017**, *851*, 012033. [CrossRef]
28. Foley, K.; Seifert, N.; Velamala, J.B.; Bennett, W.G.; Gupta, S.; Gupta, S. IRT: A modeling system for single event upset analysis that captures charge sharing effects. In Proceedings of the 2014 IEEE International Reliability Physics Symposium, Waikoloa, HI, USA, 1–5 June 2014; pp. 5F.1.1–5F.1.9.
29. Reed, R.A.; Weller, R.A.; Mendenhall, M.H.; Fleetwood, D.M.; Warren, K.M.; Sierawski, B.D.; King, M.P.; Schrimpf, R.D.; Auden, E.C. Physical Processes and Applications of the Monte Carlo Radiative Energy Deposition (MRED) Code. *IEEE Trans. Nucl. Sci.* **2015**, *62*, 1441–1461. [CrossRef]
30. Geant4-Collaboration. *Book for Application Developers, Rev 3.1 ed*; CERN: Geneva, Switzerland, 2019; p. 433.
31. Bertini, H.W. Intranuclear-Cascade Calculation of the Secondary Nucleon Spectra from Nucleon-Nucleus Interactions in the Energy Range 340 to 2900 MeV and Comparisons with Experiment. *Phys. Rev.* **1969**, *188*, 1711–1730. [CrossRef]
32. Folger, G.; Ivanchenko, V.N.; Wellisch, J.P. The Binary Cascade. *Eur. Phys. J. A* **2004**, *21*, 407–417. [CrossRef]
33. Wright, D.H.; Koi, T.; Folger, G.; Ivanchenko, V.; Kossov, M.; Starkov, N.; Heikkinen, A.; Wellisch, H. Low and High Energy Modeling in Geant4. In Proceedings of the Fourth Huntsville Gamma-Ray Burst Symposium; AIP Publishing: Melville, NY, USA, 2007; Volume 896, pp. 11–20.
34. Wright, D.H.; Kelsey, M.H. The Geant4 Bertini Cascade. *Nuclear Instruments and Methods. Accel. Spectrometers Detect. Assoc. Equip.* **2015**, *804*, 175–188. [CrossRef]
35. Koning, A.; Rochman, D. Modern Nuclear Data Evaluation with the TALYS Code System. *Nucl. Data Sheets* **2012**, *113*, 2841–2934. [CrossRef]
36. Otuka, N.; Dupont, E.; Semkova, V.; Pritychenko, B.; Blokhin, A.; Aikawa, M.; Babykina, S.; Bossant, M.; Chen, G.; Dunaeva, S.; et al. Towards a More Complete and Accurate Experimental Nuclear Reaction Data Library (EXFOR): International Collaboration Between Nuclear Reaction Data Centres (NRDC). *Nucl. Data Sheets* **2014**, *120*, 272–276. [CrossRef]
37. Agency, N.E. Joint Evaluated Fission and Fusion (JEFF) Nuclear Data Library. 6 March 2019. Available online: https://www.oecd-nea.org/dbdata/jeff/ (accessed on 23 July 2019).
38. Horiwitz, Y.S.; Dubi, A. A proposed modification of Burlin's general cavity theory for photons. *Physics Med. Biol.* **1982**, *27*, 867–870. [CrossRef]
39. Inguimbert, C.; Duzellier, S. SEU rate calculation with GEANT4 (comparison with CREME 86). *IEEE Trans. Nucl. Sci.* **2004**, *51*, 2805–2810. [CrossRef]
40. Schwank, J.; Shaneyfelt, M.; Baggio, J.; Dodd, P.; Felix, J.; Ferlet-Cavrois, V.; Paillet, P.; Lambert, D.; Sexton, F.; Hash, G.; et al. Effects of particle energy on proton-induced single-event latchup. *IEEE Trans. Nucl. Sci.* **2005**, *52*, 2622–2629. [CrossRef]
41. Raine, M.; Jay, A.; Richard, N.; Goiffon, V.; Girard, S.; Gaillardin, M.; Paillet, P. Simulation of Single Particle Displacement Damage in Silicon—Part I: Global Approach and Primary Interaction Simulation. *IEEE Trans. Nucl. Sci.* **2016**, *64*, 133–140. [CrossRef]

42. Jay, A.; Raine, M.; Richard, N.; Mousseau, N.; Goiffon, V.; Hemeryck, A.; Magnan, P. Simulation of Single Particle Displacement Damage in Silicon–Part II: Generation and Long-Time Relaxation of Damage Structure. *IEEE Trans. Nucl. Sci.* **2017**, *64*, 141–148. [CrossRef]
43. Jay, A.; Hemeryck, A.; Richard, N.; Martin-Samos, L.; Raine, M.; Le Roch, A.; Mousseau, N.; Goiffon, V.; Paillet, P.; Gaillardin, M.; et al. Simulation of Single-Particle Displacement Damage in Silicon—Part III: First Principle Characterization of Defect Properties. *IEEE Trans. Nucl. Sci.* **2018**, *65*, 724–731. [CrossRef]
44. McNulty, P.J.; Farrell, G.E.; Tucker, W.P. Proton-Induced Nuclear Reactions in Silicon. *IEEE Trans. Nucl. Sci.* **1981**, *28*, 4007–4012. [CrossRef]
45. Bagatin, M.; Gerardin, S.; Paccagnella, A.; Visconti, A.; Virtanen, A.; Kettunen, H.; Costantino, A.; Ferlet-Cavrois, V.; Zadeh, A. Single Event Upsets Induced by Direct Ionization from Low-Energy Protons in Floating Gate Cells. *IEEE Trans. Nucl. Sci.* **2016**, *64*, 464–470. [CrossRef]

Article

GaN-Based Readout Circuit System for Reliable Prompt Gamma Imaging in Proton Therapy

Vimal Kant Pandey [1,2], Cherming Tan [1,2,3,4,*] and Vivek Sangwan [1]

1. Center for Reliability Science and Technology, Chang Gung University, Wenhua 1st Road, Guishan Dist., Taoyuan City 33302, Taiwan; vimalpandey94@gmail.com (V.K.P.); sangwanvivek81@gmail.com (V.S.)
2. Department of Electronic Engineering, Chang Gung University, Wenhua 1st Rd., Guishan Dist., Taoyuan City 33302, Taiwan
3. Center for Reliability Engineering, Ming Chi University of Technology, New Taipei City 24301, Taiwan
4. Department of Urology, Chang Gung Memorial Hospital, Guishan, Taoyuan City 33302, Taiwan
* Correspondence: cherming@ieee.org; Tel.: +886-3-2118-800-3872

Abstract: Prompt gamma imaging is one of the emerging techniques used in proton therapy for in-vivo range verification. Prompt gamma signals are generated during therapy due to the nuclear interaction between beam particles and nuclei of the tissue that is detected and processed in order to obtain the position and energy of the event so that the benefits of Bragg's peak can be fully utilized. This work aims to develop a gallium nitride (GaN)-based readout system for position-sensitive detectors. An operational amplifier is the module most used in such a system to process the detector signal, and a GaN-based operational amplifier (OPA) is designed and simulated in LTSpice. The designed circuit had an open-loop gain of 70 dB and a unity gain frequency of 34 MHz. The slew rate of OPA was 20 V/µs and common mode rejection ratio was 84.2 dB. A simulation model of the readout circuit system using the GaN-based operational amplifier was also designed, and the result showed that the system can successfully process the prompt gamma signals. Due to the radiation hardness of GaN devices, the readout circuit system is expected to be more reliable than its silicon counterpart.

Keywords: GaN; operational amplifier; proton therapy; prompt gamma imaging

Citation: Pandey, V.K.; Tan, C.M.; Sangwan, V. GaN-Based Readout Circuit System for Reliable Prompt Gamma Imaging in Proton Therapy. *Appl. Sci.* **2021**, *11*, 5606. https://doi.org/10.3390/app11125606

Academic Editor: Frank Walther

Received: 27 April 2021
Accepted: 7 June 2021
Published: 17 June 2021

Publisher's Note: MDPI stays neutral with regard to jurisdictional claims in published maps and institutional affiliations.

Copyright: © 2021 by the authors. Licensee MDPI, Basel, Switzerland. This article is an open access article distributed under the terms and conditions of the Creative Commons Attribution (CC BY) license (https://creativecommons.org/licenses/by/4.0/).

1. Introduction

Proton therapy has a unique feature that makes it more advantageous compared to conventional radiotherapy. This unique feature stems from its dose distribution which remains quasi-constant at a low level along the path traveled by proton and rises sharply to a maximum value at Bragg's peak towards the end. Beyond Bragg's peak, the dose absorbed is almost negligible, reducing the risk of damage to the healthy tissues [1–3]. To fully leverage this Bragg's peak, accurate in-vivo range verification is essential.

Two main techniques are employed for this in vivo range verification which are based on the by-products of the interaction between patient and the incident irradiation. They are PET (positron emission tomography) which uses the delayed gamma [4] and prompt gamma imaging which uses prompt gamma [5]. Prompt gamma signals can easily penetrate through the tissues and emerges without much interaction with the tissues [4], hence the information of the location of its origin is not distorted, renders accurate identification of the Bragg's peak position [6–8], and thus it is commonly used.

Silicon photomultipliers (SiPM) and position-sensitive multi-output detectors are commonly employed for detecting the prompt gamma signals. The detector commonly consists of a multi anode position-sensitive photomultiplier tube (PSPMT) which has 8 × 8 multi-anode spaced with a pixel size of 6 mm × 6 mm per anode [7]. A scintillator crystal is used along with the detector to generate fluorescence when a prompt gamma signal is incident on it. This light is then converted into photoelectrons on the photocathode of the

PSPMT and a current signal is generated. The PSPMT current signals are conventionally processed by a preamplifier and a shaping circuit, and the shaped signals are converted into digital signals using analog to digital converters. A special algorithm is used to extract the energy and position of the radiation from the digital signals [9,10]. Conventionally, these circuits that process the PSPMT signals are built using an operational amplifier (OPA), and they must be placed in very close proximity to the detector to reduce signal loss and distortion, and hence their operation environment is radiative.

In proton therapy, secondary neutrons are also generated along with the prompt gamma signals. The main source of secondary neutrons is the treatment unit and the patients themselves. The neutrons generated from the treatment unit are known as the external neutrons and are generated due to the interaction of proton with beam delivery system, whereas neutrons generated from the patient is a result of proton interaction within the patient body which are known as internal neutrons [11,12]. These neutrons will negatively affect the health of the patient as well as the electronics present in the treatment room such as electronics/data acquisition systems, monitoring devices, robotic patient positioners and onboard imaging, etc. [13,14]. The radiation hardness of these electronic devices is of great concern as they are operating in the radiation environment and their durability will be compromised, as can also be observed in our treatment room. The secondary and scattered radiation field present in and around the passive proton treatment nozzle was evaluated and their results showed that the dose at any point is highly dependent on the position relative to the isocenter [12]. It was found that the dose on the gantry is approximately 3 to 6 times higher than that at the gantry floor. Since the readout circuits are in close proximity to the isocenter, their reliability and performance are greatly affected by the neutrons.

Several studies [15–21] have been conducted recently on the commercial OPA Integrated Circuits (ICs) under neutron radiation and they found that the characteristics of OPAs such as gain, slew rate, offset current, etc. are degraded after the radiation. The most used OPA, µA741 was studied under neutron and gamma rays, and results showed that the frequency behavior of the IC i.e., the gain-bandwidth product and slew rate, degraded after irradiation [19]. These changes in the gain and slew rate will affect the shape of the output signal; for example, a sinusoidal signal became a sawtooth wave due to these changes. In reference [16], LM124 from three different manufactures was irradiated under 1 MeV neutron radiation at two different fluences of 1×10^{12} ncm^{-2} and 5×10^{12} ncm^{-2}, respectively. All three ICs showed a decrease in the supply bias current, open-loop gain, and slew rate, except one of the devices which showed an increase in the slew rate after irradiation, due to the increase in the base current of the transistor used in the buffer stage.

Most of the ICs mentioned above are designed using bipolar transistors, however, neutrons can also significantly affect the characteristics of the metal oxide semiconductor devices (MOS) [22–24]. Neutrons are non-ionizing particles which upon interaction with electronic devices causes displacement damage (DD) or non-ionizing energy loss (NIEL) damage [25]. Both forms of damage will affect the properties of bipolar, MOS as well as the passive components [25]. The silicon dioxide layer in Metal Oxide Semiconductor Field Effect Transistor (MOSFET) is most sensitive to neutron radiation. Neutron irradiation creates oxide trapped charges and interface traps which affect the characteristics of MOSFET significantly [22,26]. These traps resulted in neutron-induced oxide degradation in MOSFET as observed by Vaidya et al. [27].

Gallium nitride (GaN) is emerging as one of the promising technologies for harsh environments as it is more reliable under radiation as compared to the silicon counterpart [28–30]. The mean displacement energy of GaN is higher (approximately 21.3 eV) than the silicon (approximately 11.07 eV), and there is no oxide layer beneath the gate electrode of GaN High Electron Mobility Transistor (HEMT), hence it performs better in the radiation environment [31,32]. Low noise and high-speed operation are other advantages of GaN which makes it a more viable solution for designing a readout circuit for high detection rate radiation detectors. Due to its low noise characteristics, GaN has already been used for designing low noise amplifiers [33–35]. In view of the aforementioned advantages, we developed an operational amplifier (OPA) using GaN transistors and used it to design a readout circuit for prompt gamma signals. Such a GaN-based OPA is the first of its kind reported, to the best knowledge of the authors.

2. Materials and Methods

The reliability of GaN HEMT is superior than Si MOSFET in a neutron radiation environment as proven experimentally in our previous work in [36]. Thus, one can assure the reliability of the GaN Operational Amplifier (OPAMP) designed in this work. In this work, we chose GaN HEMT transistors from efficient power conversion (EPC) in order to design the operational amplifier and readout circuit because they show better performance in a radiation environment, as reported in the literature, where negligible variations in the GaN HEMT parameters were reported in a fast neutron ambient (up to a fluence of 1×10^{15}) [37–39].

OPAs are the most used devices for designing the various modules of the readout system as shown in Figure 1 which is modified from reference [9], including transimpedance amplifiers for current to voltage conversion, adders for signal regrouping and integrators for charge to time conversion. The operational amplifier required to design these circuits should have high gain, higher bandwidth, and high speed so that it can process the short-rise time and weak signals obtained from the photomultiplier tube (PMT). The first block of the system is a scintillator crystal along with PSPMT for detecting the prompt signal and generating the corresponding current signal. The PSPMT has 64 readout channels which is reduced to 4 channels (I1, I2, I3, and I4) by the discretized position circuit (DPC) proposed in [40]. The four current signals from the PSPMT are then converted into four voltages (V1, V2, V3, and V4) by four transimpedance amplifiers (TIAs). The output voltage of the TIA is proportional to the input current i.e.,

$$V = I \cdot R \tag{1}$$

where I is the input current to the TIA and R is the feedback resistance in the TIA circuit. The signals obtained from the TIAs are then fed to the signal regrouping block where they are combined to provide the following:

$$V_{S1} = V1 + V2 + V3 + V4 \tag{2}$$

$$V_{S2} = V1 + V2 - V3 - V4 \tag{3}$$

$$V_{S3} = V1 + V4 - V2 - V3 \tag{4}$$

The gain of the adder circuit in the signal regrouping block is set to unity so that Equations (2)–(4) can be obtained in terms of the current from the PMT as follows:

$$V_{S1} = (I1 + I2 + I3 + I4) \cdot R \tag{5}$$

$$V_{S2} = (I1 + I2 - I3 - I4) \cdot R \tag{6}$$

$$V_{S3} = (I1 + I4 - I2 - I3) \cdot R \tag{7}$$

The output signal V_{S1} is fed to the integrator and upon integration, the total charge accumulated at the detector can be obtained which provides the energy information of the radiation event. The output signals V_{S1}, V_{S2}, V_{S3}, and VIE (output of the integrator) are then sent for data processing for computing the energy and position information of the radiation event. To prevent the pulse pile-up, the output of the integrator is reset to the initial state after each detection by the reset signal provided by the data processing unit. The details of the energy and position calculation are discussed in the next section.

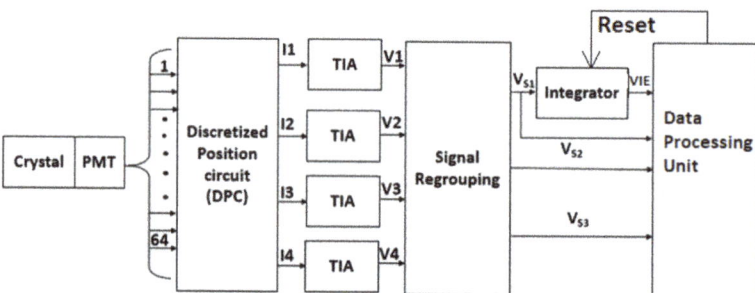

Figure 1. Signal readout block diagram.

Figure 2 shows the schematic diagram of the proposed operational amplifier. The first stage of the circuit is a dual input balanced differential amplifier formed by M1 and M2. Since a complimentary GaN transistor is not available, passive load is used throughout the design. Transistors M3, M4, M7, M9 and resistor R3 are in current mirror configuration to provide the tail current of the first, second and third stage of the circuit. The second stage is the dual input unbalanced output differential amplifier formed by M5 and M6. M10 and M11 are configured to a common source with degeneration mode and cascaded together to achieve high open-loop gain. M8 is used in common drain configuration between the second stage and the common source stage to prevent loading. Capacitor C1 is added to improve the stability of the circuit. LTSpice is employed for the design and simulation of the OPA by importing the Simulation Program with Integrated Circuit Emphasis (SPICE) model of GaN HEMT provided by EPC.

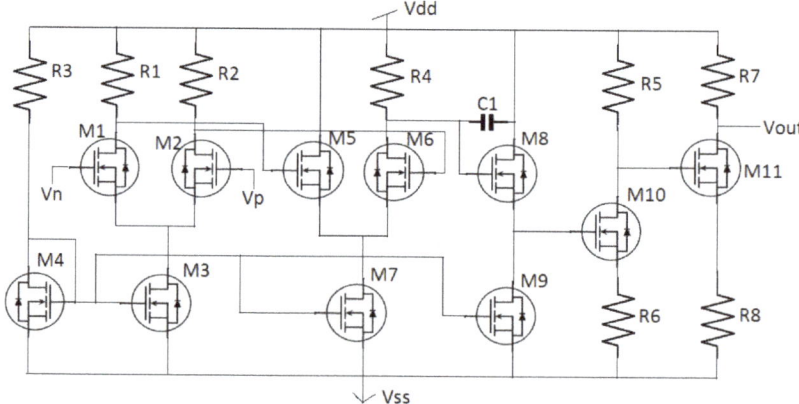

Figure 2. Proposed gallium nitride (GaN) operational amplifier. Exact values of the components cannot be shown here due to confidentiality pertaining to patent filing.

3. Results and Discussion

The open loop gain of the proposed operational amplifier circuit is obtained by AC simulation in LTSpice. From Figure 3, the open loop gain of the proposed circuit is 70 dB and the unity gain frequency is approximately 34 MHz.

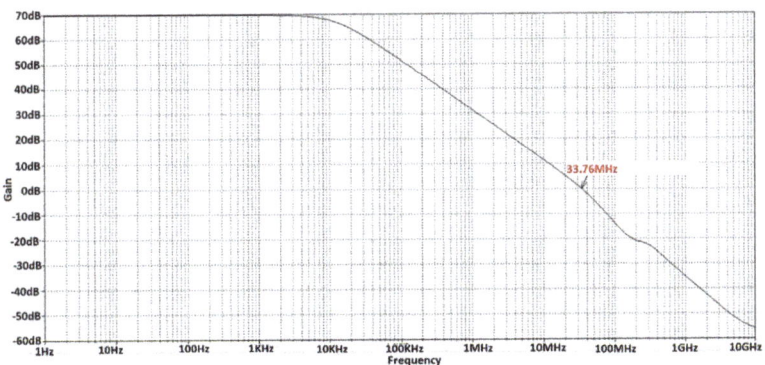

Figure 3. Open loop gain of proposed operational amplifier having a unity gain frequency of 33.76 MHz.

To compute the common mode rejection ratio (CMRR) of the OPA, a simulation setup shown in Figure 4a is designed and a sine wave of 2 V peak-to-peak is applied to the input terminal. Figure 4b shows the simulation result, the output peak-to-peak voltage obtained is 619 mV. CMRR is calculated by using the following equation:

$$\text{CMRR} = \left(1 + \frac{R_F}{R_1}\right) \cdot \left(\frac{Vin(\text{peak}-\text{to}-\text{peak})}{Vout(\text{peak}-\text{to}-\text{peak})}\right) \tag{8}$$

On putting the values in Equation (8), we obtain:

$$\text{CMRR} = \left(1 + \frac{560\text{k}\Omega}{100\Omega}\right) \cdot \left(\frac{2\text{V}}{619\text{mV}}\right) = 16158.3$$

$$\text{CMRR in dB} = 20\log(16158.3) = 84.2\text{dB} \tag{9}$$

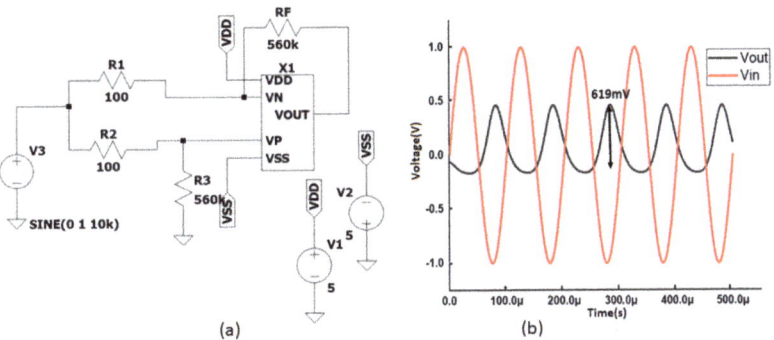

Figure 4. (a) Common mode rejection ratio (CMRR) simulation setup. (b) Simulation output.

Slew rate is another important factor of the OPA which can be measured by applying a step signal at the input of OPA and the rate of change of the signal from 10% to 90% at the output of OPA is measured [41]. For computing the slew rate, the proposed OPA is configured in unity gain mode and an ideal pulse of 1 V amplitude is applied to input terminal as shown in Figure 5a. The corresponding simulation output is shown Figure 5b. Slew rate is computed by the following equation:

$$\text{Slew rate} = \frac{\Delta V_{out}}{\Delta time} = \frac{(V_{out90\%} - V_{out10\%})}{(t_{90\%} - t_{10\%})} \tag{10}$$

On putting the simulated value from Figure 5b in Equation (10), the slew rate obtained is given as

$$\text{Slew rate} = \frac{(898.96\text{mV} - 97.92\text{mV})}{(1.045\mu s - 1.005\mu s)} = 20.03 \, \frac{\text{V}}{\mu s} \quad (11)$$

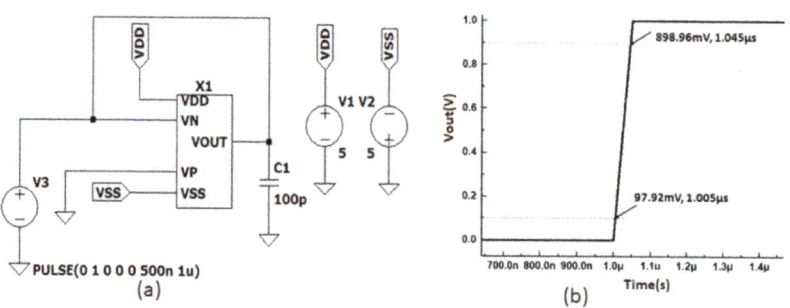

Figure 5. (a) Slew rate simulation model. (b) Simulation output.

3.1. Operational Amplifier (OPA) as Current to Voltage Converter

To convert the current from PMT into voltage, four transimpedance amplifiers are required as shown in Figure 2. The proposed OPA is configured as transimpedance amplifier as shown in Figure 6a. Resistor R1 is used as feedback resistor and the output voltage of the circuit is given by:

$$Vout = I1 \cdot R1 \quad (12)$$

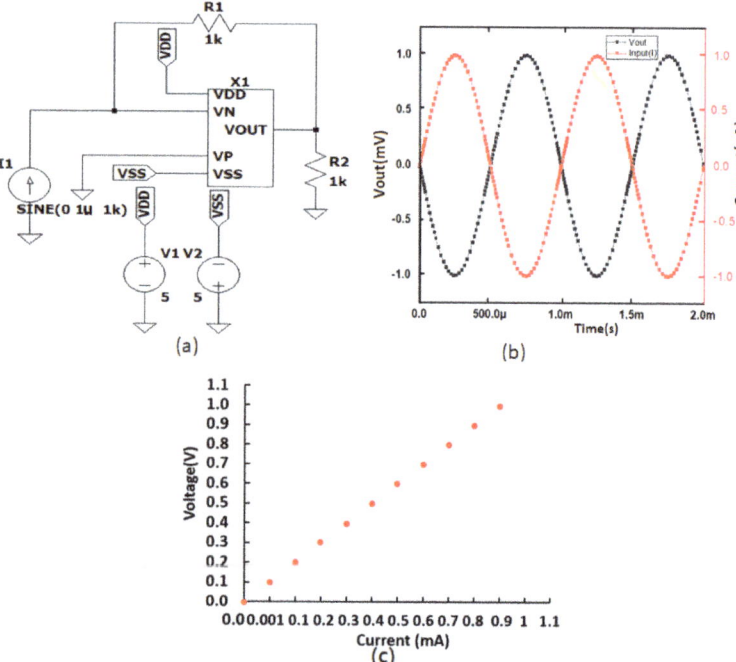

Figure 6. (a) Operational amplifier (OPA) as transimpedance amplifier (b) Simulation result (c) Simulation results showing the Linearity of TIA for 1 μA to 1 mA input current.

Figure 6b is the simulation result of the transimpedance amplifier, and it shows that, for the applied input current of 1 µA the corresponding output voltage is 1 mV, according to that given by Equation (9). The current from the PSPMT is in the range of tens of µA and Figure 6c shows the linearity of the TIA for the current range from 1 µA to 1 mA.

3.2. OPA as Adder

As given in Equations (2)–(4), an adder circuit is required for signal regrouping. The proposed OPA is configured as an adder circuit as shown in Figure 7a. Figure 7b shows the transient simulation result of the circuit. The output voltage of the circuit is the sum of the applied voltages V1, V2, and V3. The gain of the circuit is set to unity by choosing all the resistors of the same value. The amplitude of the applied input voltages is 1 mV, 2 mV, and 3 mV, respectively, and the output voltage is 6 mV as can be seen from Figure 7b.

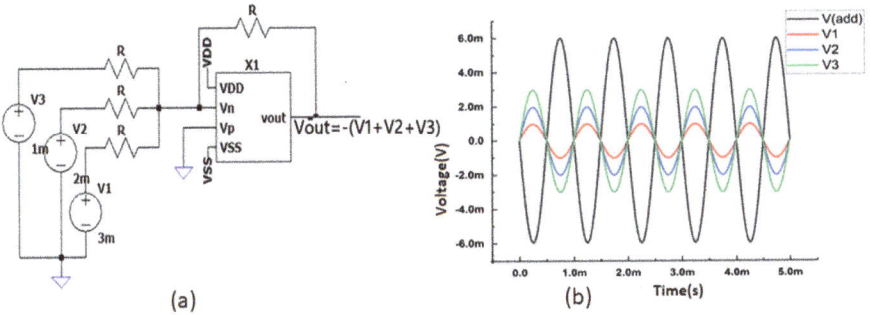

Figure 7. (**a**) Adder circuit. (**b**) Simulation output.

3.3. OPA as Integrator

In the readout circuit, integrators are commonly used for pulse shaping or charge-to-time conversion. Our proposed OPA is configured as an integrator as shown in Figure 8a, and the corresponding simulation result is shown in Figure 8b. It can be seen from Figure 8b that the square wave input signal is converted into an approximated ramp signal at the output, verifying the operation of the integrator circuit. A perfect ramp signal cannot be obtained due to the exponential charging and discharging characteristic of capacitor C1.

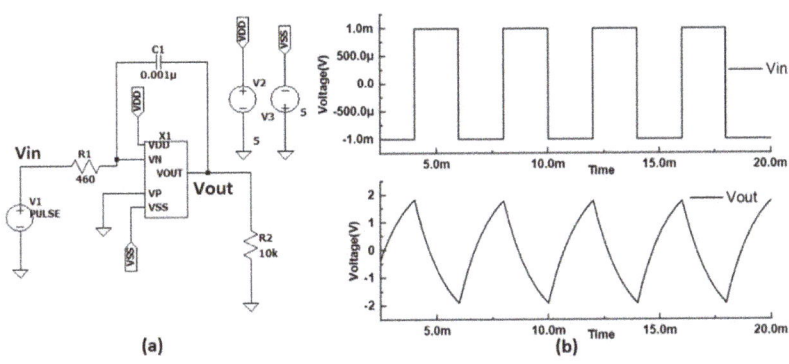

Figure 8. (**a**) Integrator circuit. (**b**) Simulation output.

3.4. Gallium Nitride (GaN) Readout Circuit for Position-Sensitive Photomultiplier Tube (PSPMT)

With the various modules designed using our proposed OPA and verified, the complete GaN-based readout circuit is built as shown in Figure 9. Block X1 is discretized position circuit; X2, X3, X4, X5 are the transimpedance amplifiers; X6 is the adder; X7, and

X8 are the adder-subtractors. The outputs from the adder and adder-subtractor circuits VS1, VS2, and VS3 will be sent to the data processing unit as shown in Figure 1 for computing the position of the radiation event. The output VS1 is also fed to the input of integrator whose output VIE will be fed to the data processing unit for computing the energy of the radiation event.

To verify the behavior of the readout circuit, the current obtained from the SiPM due to radiation interaction is mathematically modeled in LTSpice. The PM can be represented by a current source in parallel to a capacitor and the current is given by the Equation (11) [42,43]:

$$I_s = \frac{Q_e \times N_{Ph} \times PDE \times G}{\tau_d - \tau_r} \left(e^{-\frac{t}{\tau_r}} - e^{-\frac{t}{\tau_d}} \right) \tag{13}$$

where Q_e is the charge of electron, G is the gain of PMT, PDE is photon detection efficiency of the scintillation crystal, N_{Ph} is the number of the photons detected, τ_r and τ_d are the rise and decay time of the scintillator. The value of the parameters in Equation (13) for S11828-3344M SiPM and CsI(Tl) crystal is given in Table 1 [43]. Figure 10 shows the equivalent circuit of PMT and the corresponding current pulse generated by Equation (13), and in our case, the current pulse is 40 µA.

This pulse was applied randomly to different channels of DPC mimicking random radiation events, and the corresponding current signal (I1, I2, I3, I4) obtained from the four channels of the DPC is shown in Figure 11a. These current signals are then converted into voltage signals V1, V2, V3 and V4 by the TIAs shown in Figure 11b. The simulation result shows that the output signal obtained from the TIA has minimal shape distortion and follows Equation (1).

Figure 12 shows the outputs of adder X6, and integrator, respectively. Figure 12 shows that the reset signal remains low until the integration is performed, after that it toggles to the high stage and discharges the capacitor so that the integrator returns to initial state and the next signal will be integrated. The outputs VS1 is the sum of the input currents obtained from DPC as given in Equation (5). Suppose Q1, Q2, Q3 and Q4 are the integral of currents I1, I2, I3 and I4; Q_T, which is the sum of Q1 to Q4 representing the total charge collected by the PSPMT and proportional to the energy of the prompt gamma signal.

Thus, VS1 has the information of the energy of the gamma signal. The output VIE of the integrator X9 is given as:

$$VIE = a \int VS1 dt + c \tag{14}$$

where a and c are constant. The output of the integrator is then feed to the data processing unit for extracting the energy information.

The positioning of the radiation event from the DPC circuit can be determined by the following equations [44,45]:

$$X = \frac{(A + B) - (C + D)}{A + B + C + D} \tag{15}$$

$$Y = \frac{(A + D) - (B + C)}{A + B + C + D} \tag{16}$$

where A, B, C and D represents the peak values of the signals obtained from the four corner channels of the DPC. To verify if the circuit provides the correct position information, the current pulse given by Equation (13) was injected sequentially to each of the input positions of the DPC circuit mimicking the radiation event, and 64 such simulations were independently performed. The detected position information is computed using Equations (15) and (16), where the numerator of Equation (15) is VS2, the denominator of Equation (15) is VS1, the numerator of Equation (16) is VS3, and the denominator of Equation (16) is VS1. Figure 13a shows the detected positions of the PMT arranged in an 8 × 8 array and Figure 13b shows the corresponding position coordinates. The identification of the positions is accurately determined.

Table 1. Silicon photomultiplier (SiPM) current pulse parameters.

Parameter	Value
N_{Ph}	1000
G	7.5×10^5
PDE	0.40
$\tau_d (ns)$	50
$\tau_r (\mu s)$	1

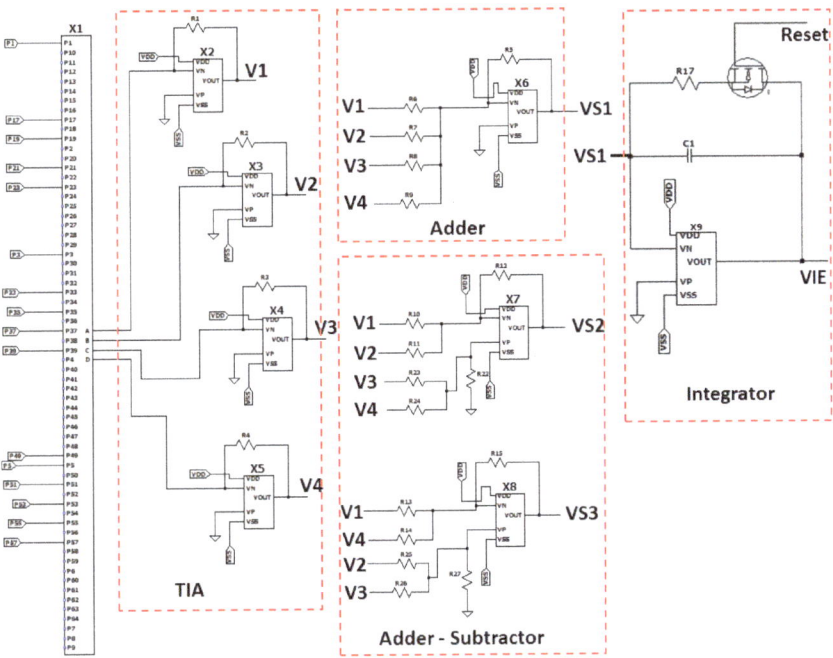

Figure 9. Simulation model of the GaN readout system.

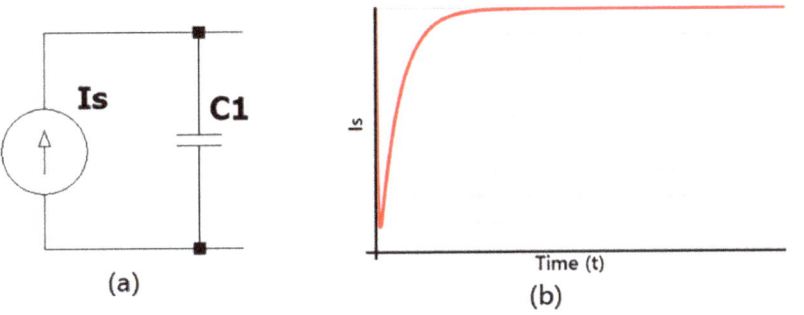

Figure 10. (a) Equivalent circuit of SiPM. (b) Current pulse generated from scintillation.

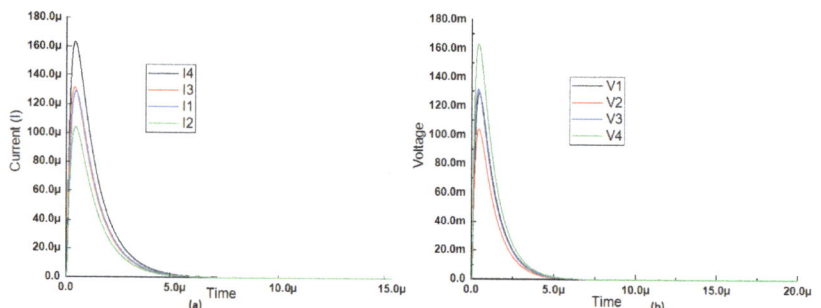

Figure 11. (**a**) Output current from four channels of discretized position circuit (DPC). (**b**) Output voltage of transimpedance amplifiers (TIAs).

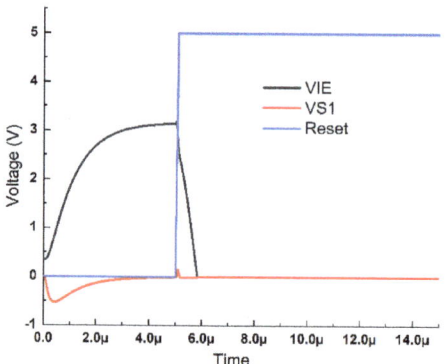

Figure 12. Simulation output of adder and integrator.

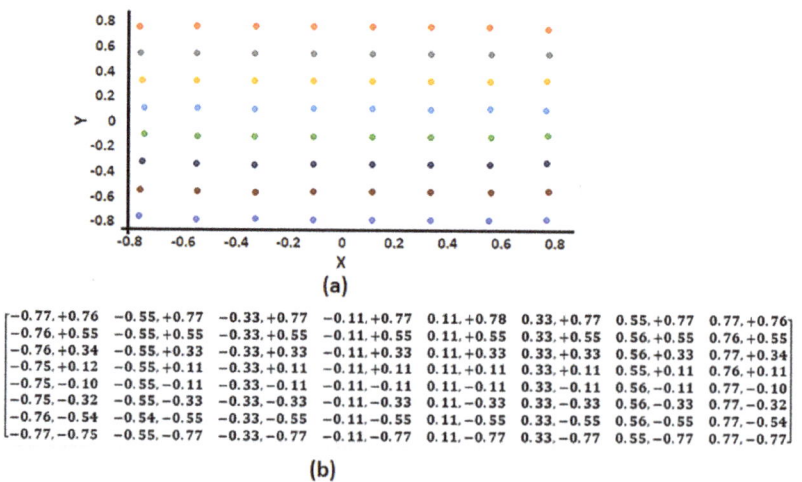

Figure 13. Position verification of the readout circuit. (**a**) position of the input current pulse applied mutual exclusively. (**b**) Computed position coordinates that correspond to the individual input current pulse.

4. Conclusions

Prompt gamma imaging is an important technique to locate the Bragg's peak position and energy. The readout circuits used for prompt gamma detection and processing are in close proximity to the radiation source. Secondary neutrons generated during proton treatment can also negatively affect the performance of the electronic devices present in the treatment room that renders their reliability important. GaN HEMT is more rad-hard compared to the silicon counterpart, and GaN-based operational amplifier is designed in this work.

Simulation results show that the designed OPA has open loop gain of 70 dB and unity gain frequency of 34 MHz. The slew rate of the OPA is 20 V/µs and common mode rejection ratio is 84.2 dB. The proposed OPA is configured for different applications such as transimpedance amplifier, integrator, and the adder which are needed in the prompt gamma readout system. Simulation results show successful operation for these applications. When these different applications are put together, a complete GaN-based prompt gamma readout circuit is implemented, and the result shows successful processing of the prompt gamma signal where its energy and position can be accurately provided for subsequent digital conversion and information extraction.

Author Contributions: Conceptualization, C.T. and V.K.P.; circuit design and simulation, V.K.P.; Writing—original draft preparation, V.K.P.; project administration, C.T.; writing—review and editing, C.T. and V.S.; All authors have read and agreed to the published version of the manuscript.

Funding: This work was funded by Chang Gung University, Grant number CIRPD2F0024.

Conflicts of Interest: All authors have no conflict of interest to the organizations mentioned in the paper.

References

1. Mukherji, A. Particle Beam Therapy: A Quick View. In *Basics of Planning and Management of Patients During Radiation Therapy*; Springer: Singapore, 2018.
2. Chuong, M.D.; Mehta, M.P.; Langen, K.; Regine, W.F. The available evidence points to benefits of proton beam therapy. *Clin. Adv. Hematol. Oncol.* **2014**, *12*, 861–864.
3. Radhe, M.; David, G. Proton Therapy—Present and Future. *Adv. Drug Deliv.* **2017**, *109*, 26–44.
4. Paganetti, H.; Fakhri, G.E.L. Monitoring proton therapy with PET. *Br. J. Radiol.* **2015**, *88*, 20150173. [CrossRef]
5. Kurosawa, S.; Kubo, H.; Ueno, K.; Kabuki, S.; Iwaki, S.; Takahashi, M.; Taniue, K.; Higashi, N.; Miuchi, K.; Tanimori, T.; et al. Prompt gamma detection for range verification in proton therapy. *Curr. Appl. Phys.* **2012**, *12*, 364–368. [CrossRef]
6. Wronska, A.; Dauvergne, D. Range verification by means of prompt-gamma detection in particle therapy. In *Radiation Detection Systems*; hal-03085504; CCSD: Charleston, SC, USA, 2020.
7. Aldawood, S.; Thirolf, P.G.; Miani, A.; Böhmer, M.; Dedes, G.; Gernhäuser, R.; Lang, C.; Liprandi, S.; Maier, L.; Marinšek, T.; et al. Development of a Compton camera for prompt-gamma medical imaging. *Radiat. Phys. Chem.* **2017**, *140*, 190–197. [CrossRef]
8. Jan, M.L.; Hsiao, I.T.; Huang, H.M. Use of a LYSO-based Compton camera for prompt gamma range verification in proton therapy. *Med. Phys.* **2017**, *44*, 6261–6269. [CrossRef] [PubMed]
9. Zhou, W.; Zhang, Z.M.; Li, D.W.; Wang, P.L.; Feng, B.T.; Huang, X.C.; Hu, T.T.; Li, X.H.; Chen, Y.; Wang, Y.J.; et al. A QTC-based signal readout for position-sensitive multi-output detectors. *Nucl. Sci. Tech.* **2016**, *27*, 1–7. [CrossRef]
10. Zhang, X.H.; Qi, Y.J.; Zhao, C.L. Design and development of compact readout electronics with silicon photomultiplier array for a compact imaging detector. *Chin. Phys. C* **2012**, *36*, 973–978. [CrossRef]
11. Aloufi, K. *Neutron Spectroscopy in Proton Therapy*; University College of London: London, UK, 2016.
12. Agosteo, S.; Birattari, C.; Caravaggio, M.; Silari, M.; Tosi, G. Secondary neutron and photon dose in proton therapy. *Radiother. Oncol.* **1998**, *48*, 293–305. [CrossRef]
13. Wroe, A.J. Evaluation and Mitigation of Secondary Dose Delivered to Electronic Systems in Proton Therapy. *Technol. Cancer Res. Treat.* **2016**, *15*, 3–11. [CrossRef]
14. Schneider, U.; Hälg, R. The impact of neutrons in clinical proton therapy. *Front. Oncol.* **2015**, *5*, 1–5. [CrossRef] [PubMed]
15. Swami, H.L.; Rathod, R.; Rao, T.S.; Abhangi, M.; Vala, S.; Danani, C.; Chaudhuri, P.; Srinivasan, R. Experimental study of neutron irradiation effect on elementary semiconductor devices using Am-Be neutron source. *Indian J. Pure Appl. Phys.* **2021**, *59*, 40–47.
16. Borel, T.; Roig, F.; Michez, A.; Azais, B.; Danzeca, S.; Roche, N.J.H.; Bezerra, F.; Calvel, P.; Dusseau, L. A typical Effect of Displacement Damage on LM124 Bipolar Integrated Circuits. *IEEE Trans. Nucl. Sci.* **2018**, *65*, 71–77. [CrossRef]
17. Franco, F.J.; Lozano, J.; Santos, J.P.; Agapito, J.A. Degradation of Instrumentation Amplifiers Due to the Nonionizing Energy Loss Damage. *IEEE Trans. Nucl. Sci.* **2003**, *50*, 2433–2440. [CrossRef]

18. Franco, F.J.; Zong, Y.; Casas-Cubillos, J.; Rodríguez-Ruiz, M.A.; Agapito, J.A. Neutron effects on short circuit currents of Op Amps and consequences. *IEEE Trans. Nucl. Sci.* **2005**, *52*, 1530–1537. [CrossRef]
19. Yan, L.; Wei, C.; Shanchao, Y.; Xiaoming, J.; Chaohui, H. Synergistic effect of mixed neutron and gamma irradiation in bipolar operational amplifier OP07. *Nucl. Instruments Methods Phys. Res. Sect. A Accel. Spectrometers Detect. Assoc. Equip.* **2016**, *831*, 334–338. [CrossRef]
20. Assaf, J. Radiation and annealing effects on integrated bipolar Operational Amplifier. *Radiat. Phys. Chem.* **2017**, *131*, 100–104. [CrossRef]
21. Boley, W.R. Compendia of TID and neutron radiation test results of selected COTS parts. In Proceedings of the 2008 IEEE Radiation Effects Data Workshop, Tucson, AZ, USA, 14–18 July 2008; pp. 142–147.
22. Amir, H.F.A.; Chee, F.P.; Salleh, S. Effects of high energy neutrons and resulting secondary charged particles on the operation of MOSFETs. In Proceedings of the 2014 International Conference on Computational Science and Technology, ICCST, Kota Kinabalu, Malaysia, 27–28 August 2014.
23. Chao, D.S.; Shih, H.Y.; Jiang, J.Y.; Huang, C.F.; Chiang, C.Y.; Ku, C.S.; Yen, C.T.; Lee, L.S.; Hsu, F.J.; Chu, K.T.; et al. Influence of displacement damage induced by neutron irradiation on effective carrier density in 4H-SiC SBDs and MOSFETs. *Jpn. J. Appl. Phys.* **2019**, *58*, SBBD08. [CrossRef]
24. Abdul Amir, H.F.; Chik, A. Neutron radiation effects on metal oxide semiconductor (MOS) devices. *Nucl. Instrum. Methods Phys. Res. Sect. B Beam Interact. Mater. Atoms* **2009**, *267*, 3032–3036. [CrossRef]
25. Makowski, D. *The Impact of Radiation on Electronic Devices with Special Consideration of Neutron and Gamma Radiation Monitoring*; Technical University of Lodz: Lodz, Poland, 2006.
26. Haider, F.A.; Chee, F.P.; Abu Hassan, H.; Saafie, S.; Afishah, A. Changes in electrical properties of MOS transistor induced by single 14 MeV neutron. In Proceedings of the AIP Conference Proceedings, Online, 22 January 2016; Volume 1704, p. 050015.
27. Vaidya, S.J.; Sharma, D.K.; Chandorkar, A.N. Neutron induced oxide degradation in MOSFET structures. In Proceedings of the International Symposium on the Physical and Failure Analysis of Integrated Circuits, IPFA 2003, Singapore, 11 July 2003; pp. 151–155.
28. Angela Chen Gallium Nitride is the Silicon of the Future. Available online: https://www.theverge.com/2018/11/1/18051974/gallium-nitride-anker-material-silicon-semiconductor-energy#:~{}:text=GaNhasawiderband,onGaNinpowerelectronics (accessed on 11 April 2021).
29. Where is GaN Going? Available online: https://epc-co.com/epc/GalliumNitride/whereisgangoing.aspx (accessed on 12 April 2021).
30. Scharf, A. *Gallium Nitride is Moving Forward*; Power Electronics Europe: Munich, Germany, 2016.
31. Zafrani, M. *Radiation Performance of Enhancement-Mode Gallium Nitride Power Devices*; EE Power: Hatfield, UK, 2020; pp. 40–42.
32. Hazdra, P.; Popelka, S. Radiation resistance of wide-bandgap semiconductor power transistors. *Phys. Status Solidi Appl. Mater. Sci.* **2017**, *214*, 1–8. [CrossRef]
33. Cha, S.; Chung, Y.H.; Wojtowwicz, M.; Smorchkova, I.; Allen, B.R.; Yang, J.M.; Kagiwada, R. Wideband AlGaN/GaN HEMT low noise amplifier for highly survivable receiver electronics. In Proceedings of the IEEE MTT-S International Microwave Symposium Digest, Fort Worth, TX, USA, 6–11 June 2004; pp. 829–831.
34. Pengelly, R.; Sheppard, S.; Smith, T.; Pribble, B.; Wood, S.; Platis, C. Commercial Gan devices for switching and low noise applications. In Proceedings of the 2011 International Conference on Compound Semiconductor Manufacturing Technology, CS MANTECH 2011, Palm Springs, CA, USA, 16–19 May 2011; pp. 27–30.
35. Helali, A.; Gassoumi, M.; Gassoumi, M.; Maaref, H. Design and Optimization of LNA Amplifier Based on HEMT GaN for X-Band Wireless-Communication and IoT Applications. *Silicon* **2020**, 1–9. [CrossRef]
36. Pandey, V.K.; Tan, C.M. Application of Gallium Nitride Technology in Particle Therapy Imaging. *IEEE Trans. Nucl. Sci.* **2021**, *68*, 1319–1324. [CrossRef]
37. Zafrani, M.; Lidow, A. Radiation Performance of Enhancement-Mode Gallium Nitride Power Devices. Available online: https://epc-co.com/epc/EventsandNews/News/ArtMID/1627/ArticleID/2933/Radiation-Performance-of-Enhancement-Mode-Gallium-Nitride-Power-Devices.aspx (accessed on 26 April 2021).
38. Lidow, A.; Smalley, K. Radiation Tolerant Enhancement Mode Gallium Nitride (eGaN ®) FET Characteristics. In Proceedings of the GOMAC Tech Conference, Las Vegas, NV, USA, 19–22 March 2012.
39. Lidow, A.; Nakata, A.; Rearwin, M.; Strydom, J.; Zafrani, A.M. Single-event and radiation effect on enhancement mode gallium nitride FETs. In Proceedings of the IEEE Radiation Effects Data Workshop, Paris, France, 14–18 July 2014; pp. 1–7.
40. Siegel, S.; Silverman, R.W.; Shao, Y.; Cherry, S.R. Simple charge division readouts for imaging scintillator arrays using a multi-channel PMT. *IEEE Trans. Nucl. Sci.* **1996**, *43*, 1634–1641. [CrossRef]
41. Slew Rate. Available online: https://training.ti.com/system/files/docs/1221-SlewRate1-slides.pdf (accessed on 22 March 2021).
42. Seifert, S.; Van Dam, H.T.; Huizenga, J.; Vinke, R.; Dendooven, P.; Löhner, H.; Schaart, D.R. Simulation of silicon photomultiplier signals. *IEEE Trans. Nucl. Sci.* **2009**, *56*, 3726–3733. [CrossRef]
43. Massari, R.; Soluri, A.; Caputo, D.; Ronchi, S. Low power readout circuits for large area silicon photomultiplier array. In Proceedings of the 6th International Workshop on Advances in Sensors and Interfaces (IWASI), Gallipoli, Italy, 18–19 June 2015; pp. 158–162.

44. Olcott, P.D.; Talcott, J.A.; Levin, C.S.; Habte, F.; Foudray, A.M.K. Compact readout electronics for Position Sensitive Photomultiplier Tubes. In *Proceedings of the IEEE Nuclear Science Symposium Conference Record*; IEEE: Piscataway, NJ, USA, 2003; Volume 3, pp. 1962–1966.
45. Jeon, S.J.; Kim, J.; Ji, M.G.; Park, J.H.; Choi, Y.W. Position Error Correction Using Homography in Discretized Positioning Circuit for Gamma-Ray Imaging Detection System. *IEEE Trans. Nucl. Sci.* **2017**, *64*, 816–819. [CrossRef]

Article

Failure Analysis of SAC305 Ball Grid Array Solder Joint at Extremely Cryogenic Temperature

Yanruoyue Li [1], Guicui Fu [2], Bo Wan [2,*], Maogong Jiang [2], Weifang Zhang [2] and Xiaojun Yan [1]

1. School of Energy and Power Engineering, Beihang University, Beijing 100191, China; lyry2011@buaa.edu.cn (Y.L.); yanxiaojun@buaa.edu.cn (X.Y.)
2. School of Reliability and Systems Engineering, Beihang University, Beijing 100191, China; fuguicui@buaa.edu.cn (G.F.); maogong@buaa.edu.cn (M.J.); 08590@buaa.edu.cn (W.Z.)
* Correspondence: wanbo@buaa.edu.cn; Tel.: +86-156-5292-8449

Received: 11 February 2020; Accepted: 9 March 2020; Published: 12 March 2020

Featured Application: It is an important trend that Pb-free materials in electronic devices/components are used in the aerospace field. Several reliability issues associated with this kind of material characteristic occurred and need in-depth studies. One of the issues is that Sn-rich material's characteristic changes under low-temperature. This change may cause components/device failure and has an influence on their reliability. This work of failure analysis provides an actual case of electronic components with failure when under harsh environments, such as extreme cryogenic temperature in space. It makes reliable use of Pb-free material under extremely cryogenic temperature conditions to be taken seriously, brings up an analysis process for this kind of failure, and suggestions for operating temperatures are put forward.

Abstract: To verify the reliability of a typical Pb-free circuit board applied for space exploration, five circuits were put into low temperature and shock test. However, after the test, memories on all five circuits were out of function. To investigate the cause of the failure, a series of methods for failure analysis was carried out, including X-ray detection, cross-section analysis, Scanning Electron Microscope (SEM) analysis, and contrast test. Through failure analysis, the failure was located in the Pb-free (Sn-3.0Ag-0.5Cu) solder joint, and we confirmed that the failure occurred because of the low temperature and change of fracture characteristic of Sn-3.0Ag-0.5Cu (SAC305). A verification test was conducted to verify the failure mechanism. Through analyzing data and fracture surface morphology, the cause of failure was ascertained. At low temperature, the fracture characteristic of SAC305 changed from ductileness to brittleness. The crack occurred at solder joints because of stress loaded by shock test. When the crack reached a specific length, the failure occurred. The temperature of the material's characteristic change was −70–−80 °C. It could be a reference for Pb-free circuit board use in a space environment.

Keywords: SAC305; BGA; low temperature; fracture failure

1. Introduction

Because of the importance of space exploration, more and more institutions are turning their research direction to the deep space. Considering the special environmental condition, there may be a variety of composite reliability issues, especially the reliability problems of electronic components operating under low-temperature conditions [1]. It is hard to maintain or replace the equipment operating in space. Once there is a problem, the consequences are difficult to predict. Therefore, it is significant to study the possible failure mechanism of electronic components in the ultra-low temperature environment [2,3].

Failure of the solder joint plays an important role in the field of electronic components' reliability [4]. For a long time, 63Sn-37Pb has become the most appropriate solder material owing to its practicability, economy, and superior performance. Research about Sn-Pb solder has been maturing, and it has been used as the main material in the packaging structure of various electronic components [4,5]. But Pb is a toxic metal and will contaminate the environment. In order to decrease the impact on the environment, the EU issued the Restriction of the Use of Certain Hazardous Substances in Electrical and Electronic Equipment (RoHS) on 13 February 2003, which accelerated the Pb-free process. However, there are some exempt applications put forward in this document [6]. These exceptions include space exploration. But with the rapid development of electronic industry and aerospace technology, traditional Sn-Pb material cannot meet the requirement of high performance and high reliability [5]. Pb-free in the space exploration field is the tendency in the future. Thus, Pb-free solder material has been developed and studied. Component manufacturers are forced to use Pb-free solder instead of 63Sn-37Pb solder [7]. Recently, Sn-Ag-Cu solder has become one of the most useful materials as a replacement of Sn-Pb solder. More studies of this kind of material have been carried out. A considerable amount of practice indicates that there are some differences between Sn-Ag-Cu (SAC) solder and Sn-Pb solder in terms of reliability [8]. With the difference in the amount of Ag and Cu, there are many kinds of Pb-free solder material, such as Sn-3.0Ag-0.5Cu (SAC305), Sn-3.8Ag-0.7Cu (SAC387), Sn-3.5Ag-0.7Cu (SAC357), and SAC with other composition. Researchers not only have made a study on the properties of basic SAC materials but also on the influence of other solder composition on material properties, such as performance of different composition and comparison of different additional quantity [9–11]. Among them, SAC305 is one of the most common Pb-free solder materials. In our actual practice, the very material of our failed Pb-free circuit solder is SAC305 as well.

Aiming at the reliability of Pb-free solder joints, there is a lot of researches and discussions. The main source of the Pb-free solder joint reliability problem is as follows: Shear fatigue and creep crack of solder joints [7,12], electro-migration [5,12], cracks formed by intermetallic compound (IMC) between solder and matrix interface [13,14], the short circuit caused by Sn whisker growth [13], and electric and chemical corrosion [15]. Based on these reliability problems, researchers, such as S. Pin [16], G Jian [17], A Surendar [18], Zijie Cai [19], and F Liu [20] have made related test studies to find out the mechanism of failure. Most of the researches are about temperature cycling, mechanical shock, and electro-migration. Xu Long [15], Liu, XG [21], and M Aamir [22] researched the influence of different material elements added in Pb-free material. In addition, X Niu [23] and D. S. Liu [24] made studies on the low temperature's effect on solder joint fracture behavior. Yet, the lowest temperature in these studies is −45 °C. Further studies are needed to explain the failure occurred under extremely cryogenic temperatures in a space environment.

The aim of this work was to find out the root cause of a failure occurred in practice. The failed component was the memory of a typical Pb-free circuit board, which is used for space application. First, failure analysis methods were used to confirm the failure mechanism. Then, in order to verify the results of failure analysis, samples were designed, as well as put into the verification test. By analyzing the results of low-temperature tensile test for SAC305 solder material, the root cause of this failure case was put forward. Finally, the conclusion was drawn at the end of the manuscript.

2. The Subject of Study and Failure Background

Because of the large number of applications of Pb-free components, the Pb-free circuit board used for deep-space applications attracts a lot of interest. At present, users want to know whether the Pb-free circuit board can be used reliably in space.

In order to verify the reliability of a typical Pb-free circuit board in aerospace applications, five circuits were put into low-temperature and shock test, which was called a qualification test in the subsequent section. Combining with application requirements and JEDEC standards (JESD22-B110 and JESD22-A119), the test temperature was −100 °C, and the test acceleration was 100G, 0.5-millisecond duration, and half-sin pulse. After the tests, all five circuits had failed, depending on the results of the printed circuit board (PCB) function test. There was no output of the circuit. Function tests for every component were conducted, and it could be confirmed that the failure of the PCB was caused by the memory (Figure 1) on the circuit. The memory could not store and read data normally. The package of the memory was Ball Grid Array (BGA), and the solder material was SAC305. Further analysis is needed in order to find out the root cause of the failure.

Figure 1. The failed memory on the board.

3. Failure Analysis of the Memory

A series of methods were used to make the failure analysis. Figure 2 shows the process of failure analysis. The appearance of all five memories was observed under a stereoscopic microscope, and there was no obvious damage on the surface. Two of the five memories were removed to put into the electronic function test. The results showed that the components themselves were intact. Thus, we guessed that the failure occurred at solder joints. When the failure was located at the BGA solder joint, X-ray detection was carried out at first. Second, the cross-section was analyzed using a metallurgical microscope. Then, the solder ball tensile test was conducted, and the fracture surface of the ball was observed using SEM. Finally, a contrast test under room temperature was conducted to find out the failure mechanism.

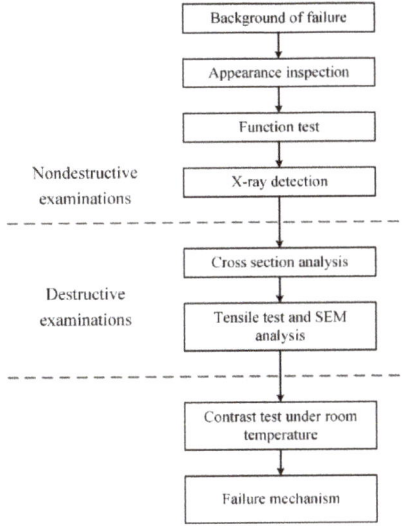

Figure 2. The process of failure analysis.

3.1. X-ray Detection

X-ray detection is one of the most important nondestructive detection methods used in failure analysis. It can find out defects of solder balls, such as voids, and size inconsistency. Figure 3 shows the BGA of the memory through X-ray. The deformation of solder balls could be seen obviously (the red circle in Figure 3). However, observation of deformation could not explain the failure cause and mechanism. Therefore, further analysis of the deformed solder balls is necessary.

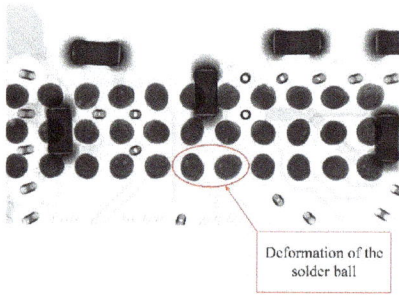

Figure 3. X-ray detection shows the deformation of the solder ball.

3.2. Cross-Section Analysis

Afterward, a cross-section of solder joints was prepared, and cracks were found under a metallurgical microscope, as shown in Figure 4. It could be seen that the solder ball on the left had a crack almost penetrating it, and the one on the right was broken completely. Combined with the previous conclusion, the memory itself was intact; it indicated that the failure was caused by cracks of the BGA solder joint.

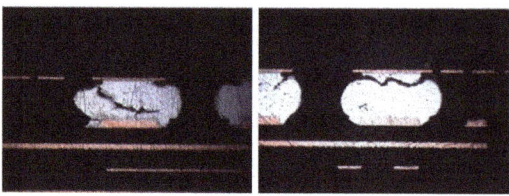

Figure 4. Cracks in the solder joint observed under a metallurgical microscope.

3.3. Tensile Test and SEM Analysis

Considering the special Pb-free material (SAC305) and the extreme environmental condition, the authors speculated that the failure was relevant to low temperature based on existing researches [25,26]. The other three memories of the failed circuit board were put into the tensile test. The universal tensile testing machine was used to pull the memory off the board. The corresponding fixture was made, and the tensile test was conducted under room temperature with a constant speed of 0.1 mm/min. The fracture occurred on the solder ball. Then, SEM was used to observe the fracture surface of the solder joints. The surface morphology under SEM is shown in Figure 5.

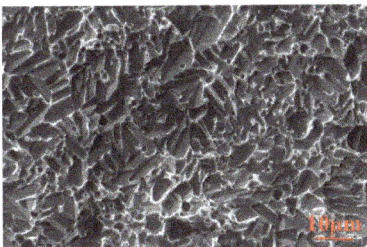

Figure 5. The surface morphology of BGA under SEM shows the brittle fracture.

From Figure 5, the obvious intergranular fracture could be seen. It is the characteristic of brittle fracture. It could be observed that there was a brittle fracture of the solder joint. Theoretically, the fracture behavior of SAC305 should be ductile at a normal temperature. Actually, the result of SEM showed that there was a change in material characteristics. According to relative studies, Sn-based solder transforms from ductile to brittle at low temperatures. This is because of an isomer transition phenomenon (commonly known as Sn-pest) of Sn-rich solder alloys when the temperature is low, and for SAC305, the temperature is lower than −30 °C [27]. The fracture changes from β-Sn to α-Sn and the volume of this kind of Sn fracture can expand by 26%, resulting in partial or total fracture of the solder joint [28]. We think the reason is that low temperature causes the change in material characteristics, and it is much easier to fracture under cryogenic and shock tests because of the brittleness. In order to verify our conjecture, a contrast test under room temperature was designed.

3.4. Contrast Test Under Room Temperature

New five identical Pb-free circuit boards were put into shock test with the same profile, but under normal temperature (25 °C) as contrasted. The test condition was 100G acceleration, 0.5-millisecond duration, 125 cm/s velocity change, and half-sine pulse. The circuit board was fixed on the test bench and subjected to a total of 12 shocks, which were two shock pulses of the peak acceleration, velocity change, and pulse duration in each of the positive and negative directions of three orthogonal axes (X, Y, and Z). All five circuit boards worked normally after the contrast test. There was no deformation and crack on solder balls, as shown in Figures 6 and 7. In the meanwhile, the tensile test and SEM analysis were also conducted, and the result is shown in Figure 8. Typical dimple fracture surface could be recognized, and it was an obvious ductile fracture surface that was different from Figure 5. The profile of the contrast test only changed the temperature compared with the initial cryogenic and shock test. Thus, we believed that cryogenic temperature made an impact on solder joints.

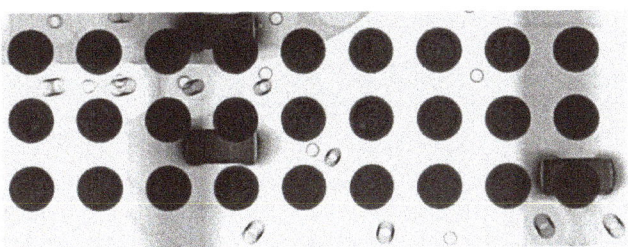

Figure 6. The X-ray detection result of an intact sample.

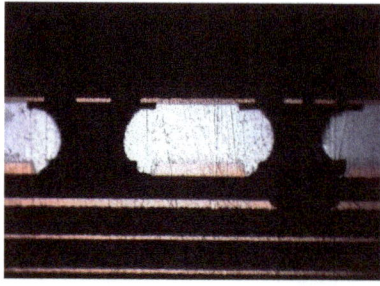

Figure 7. Intact solder joint observed under a metallurgical microscope.

Figure 8. The contrast surface morphology of BGA under SEM at room temperature shows the ductile fracture.

3.5. Failure Mechanism of BGA Solder Joints

According to the previous analysis, the mechanical properties of solder joints at low-temperature are quite different from those at room temperature. It is the characteristic of Sn-rich material. The change of properties at ultra-low temperature will greatly influence the reliability of electronic components.

Through the series of tests and analysis, the failure mechanism of the Pb-free PCB could be confirmed. At low temperature (−100 °C), the fracture behavior of SAC305 changed from ductile to brittle. Meanwhile, because of the stress caused by shock test, cracks occurred in solder balls, and finally, the failure occurred. Aiming at SAC305, there was still a remaining problem. At what temperature would SAC305 material property change? To find the transition temperature to guide actual operation, further tests need to be carried out.

4. Failure Mechanism Verification

In order to verify the failure mechanism and find the transition temperature of SAC305, a low-temperature tensile test was designed, and corresponding samples were made and experienced the test process. Then, SEM analysis was used to confirm the material characteristic.

4.1. Experiments

Through a large number of experiments, researchers have found that when the temperature gradually decreases, the ductile fracture strength of different solder material significantly increases [29]. When the temperature reaches the transition temperature range, the fracture characteristic of solders changes obviously, as well as the energy required for fracture. The ductile fracture would transform into brittle fracture [30].

According to the previous researches, and because we could judge the ductileness/brittleness through tensile test data and fracture surface morphology, a low-temperature tensile test was conducted. The experimental scheme for obtaining the cryogenic mechanical property parameters of SAC305 solder joints were designed and indicated in Table 1. Tensile tests at different temperatures were carried out by the universal material testing machine.

At every specific temperature, the immersion time of samples was 0.5 h. Afterward, the tensile test was started with 0.01/s tension rate. The values of displacement from maximum tensile stress to complete fracture and tensile strength could be recorded as representative of the mechanical property of SAC305 material.

Table 1. Test temperature and corresponding samples.

Number of Samples	Temperature (°C)
1#, 2#	25
3#, 4#	−50
5#, 6#	−70
7#, 8#	−80
9#, 10#	−100

4.2. Sample Introduction

According to the design of the low-temperature tensile test, corresponding samples of SAC305 material and clamp for the tensile test were designed. Figure 9 is the design drawing of the sample [31]. According to the experimental requirements for obtaining the low-temperature mechanical characteristics of the solder joint, Cu was selected as the base material for welding. SAC305 was used to connect the two Cu pieces. According to Figure 9, the length of the welding was 6 mm, the width was 1 mm, and the thickness of the test sample was 1 mm. This kind of design provided convenience for subsequent tests. Before welding, solder resist was applied to the surface of Cu blocks except for the weld area. Then, the two blocks were put into the clamp, and the solder paste was applied between the weld interface of two blocks. In order to simulate the actual welding process, a reflow welding temperature profile was set, and the samples were welded through the heating platform. Holding and welding temperatures were 175 °C and 250 °C, respectively. The duration of each stage could be seen in Figure 10.

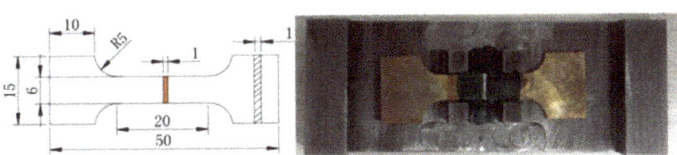

Figure 9. The design and the actual sample prepared for further tests.

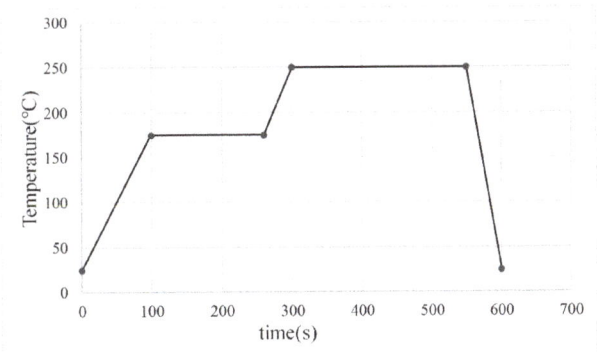

Figure 10. The welding temperature profile of sample preparation.

4.3. Results and Discussion

Through the test, the property parameters of solder joints at low temperature, the fracture surface samples, and the temperature range of ductile-brittle transition of solder joints could be obtained. Fracture strength and the displacement from the maximum tensile stress to the final fracture of Pb-free solder joints at different temperatures were obtained. Figure 11 shows the tensile strength and displacement curves of SAC305. According to the graphs and focus on the curve after the maximum tensile strength, it could be seen that the fracture mode of the solder joints was a ductile fracture at 25 °C. The slope of the curve after fracture changed slowly. With the decrease of temperature, slopes of the curve after the maximum tensile strength increased suddenly, meaning the mechanism changed from ductile fracture to brittle fracture. The transition temperature range of SAC305 solder joints was −70–−80 °C.

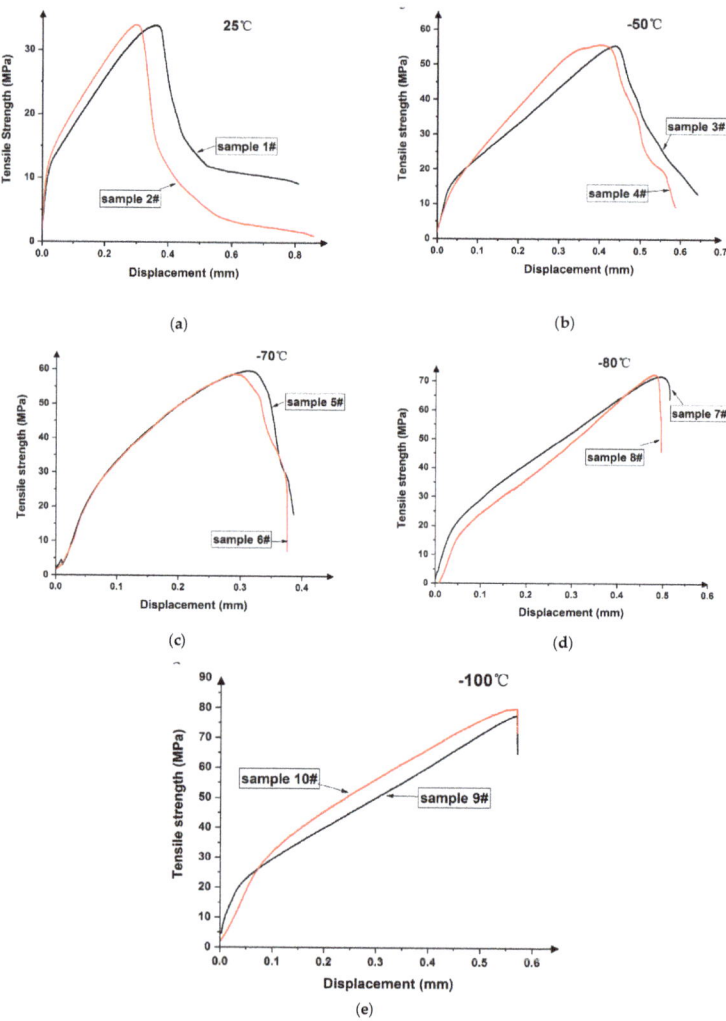

Figure 11. Tensile strength-displacement curve of Sn3.0Ag0.5Cu solder joints at different temperatures (**a**) 25 °C; (**b**) −50 °C; (**c**) −70 °C; (**d**) −80 °C; (**e**) −100 °C.

Table 2 shows the data of displacement from maximum tensile stress to complete fracture. This parameter is the characterization of the material toughness. The larger the value, the better the toughness. With the decrease in temperature, the values of displacement decreased as well. That is to say, the characteristic of ductile fracture became weaker, and the toughness of SAC305 was worse. The relationship between the toughness and the displacement from maximum tensile stress to complete fracture could be explained as: when the toughness decreased, the displacement decreased as well. The shock resistance of SAC305 gradually decreased. With the brittle fracture stage, when the external stress reached the fracture limit, small displacement could cause the fracture of the solder joint.

Table 2. Displacement from maximum tensile stress to complete fracture of Sn3.0Ag0.5Cu solder joint at different temperatures.

Temperature (°C)	Displacement from Maximum Tensile Stress to Complete Fracture (mm)		
	Sample 1	Sample 2	Average
25	0.577	0.601	0.589
−50	0.134	0.152	0.143
−70	0.082	0.076	0.079
−80	0.033	0.023	0.028
−100	0.009	0.008	0.009

Figures 12–16 are the SEM graphs of the fracture surface at different temperatures. At 25 °C, −50 °C, and −70 °C, dimples were shown in the fracture surface morphology. It was obvious that the fracture mechanism was a ductile fracture. At −80 °C and −100 °C, the figures showed the mechanical changes to brittle fracture. Different surface morphology under lower temperatures showed different material fracture characteristics from that under higher temperatures. From Figures 15 and 16, intergranular fracture surface and river pattern could be perceived. They are the characteristics of brittle fracture. The results could be a supplementary instruction of the transition of material characteristics and its temperature range.

Through the analysis of test results, we could note that at cryogenic temperature, the fraction mechanism of SAC305 material changed from ductility to brittleness. The transition temperature range was −70−−80 °C. Though the strength of extension was greater at a lower temperature, the material brittleness was higher as well. That means if there are cracks in SAC305 solder joints, which have been verified to easily occurring after the Pb-free welding process [32], less stress can cause crack growth and even fracture of Pb-free solder joint at low temperature.

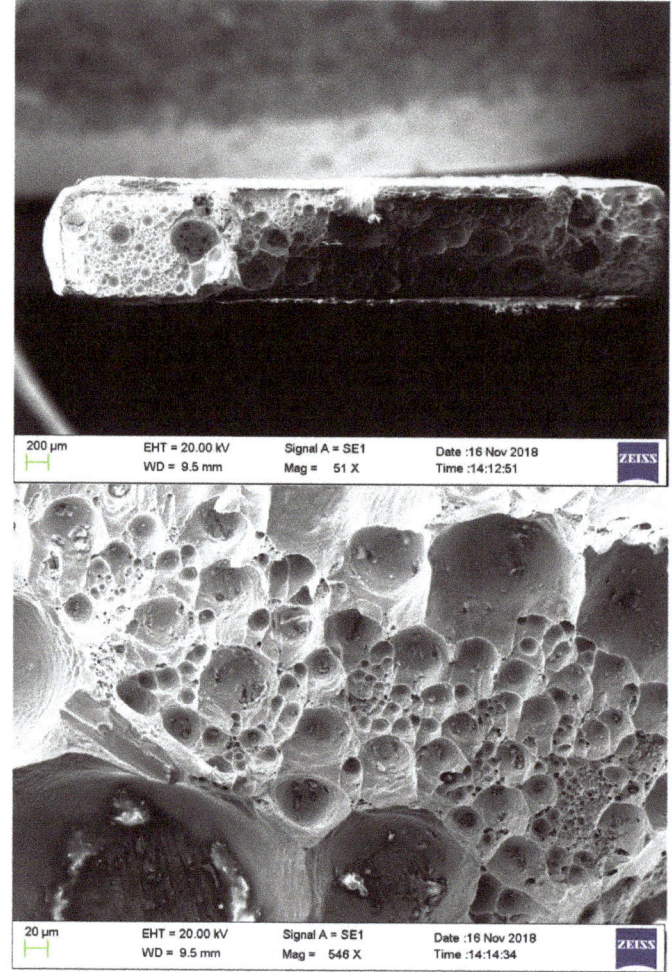

Figure 12. SEM graph of fracture surface at 25 °C (ductile).

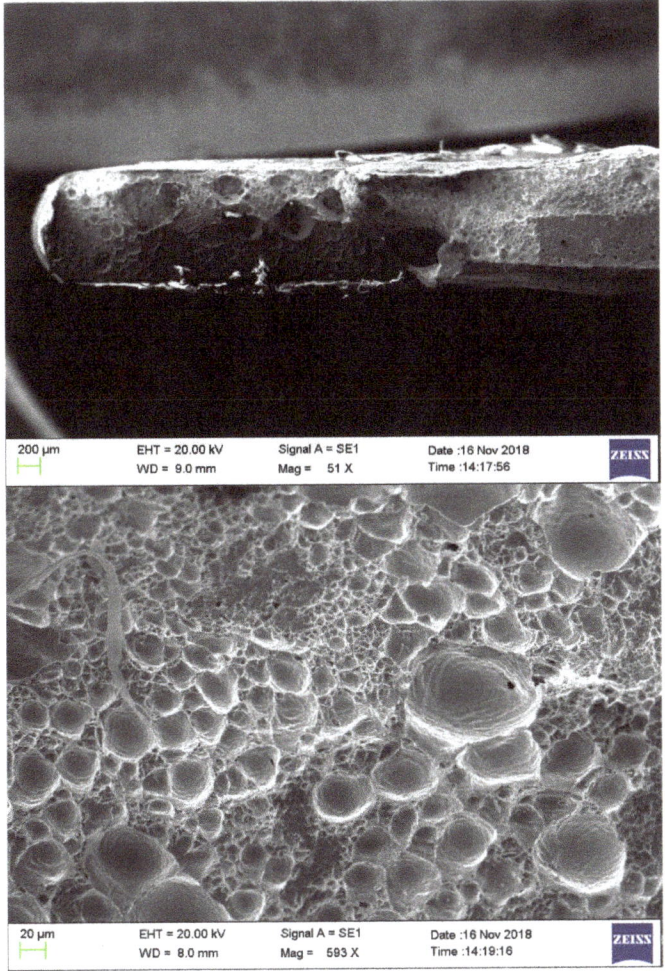

Figure 13. SEM graph of fracture surface at −50 °C (ductile).

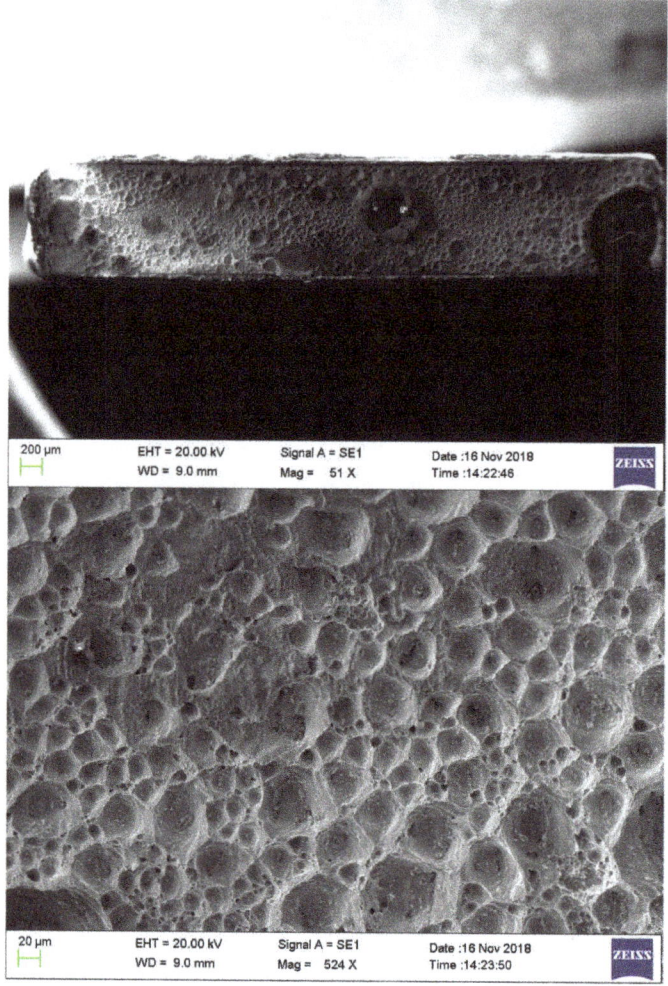

Figure 14. SEM graph of fracture surface at −70 °C (ductile).

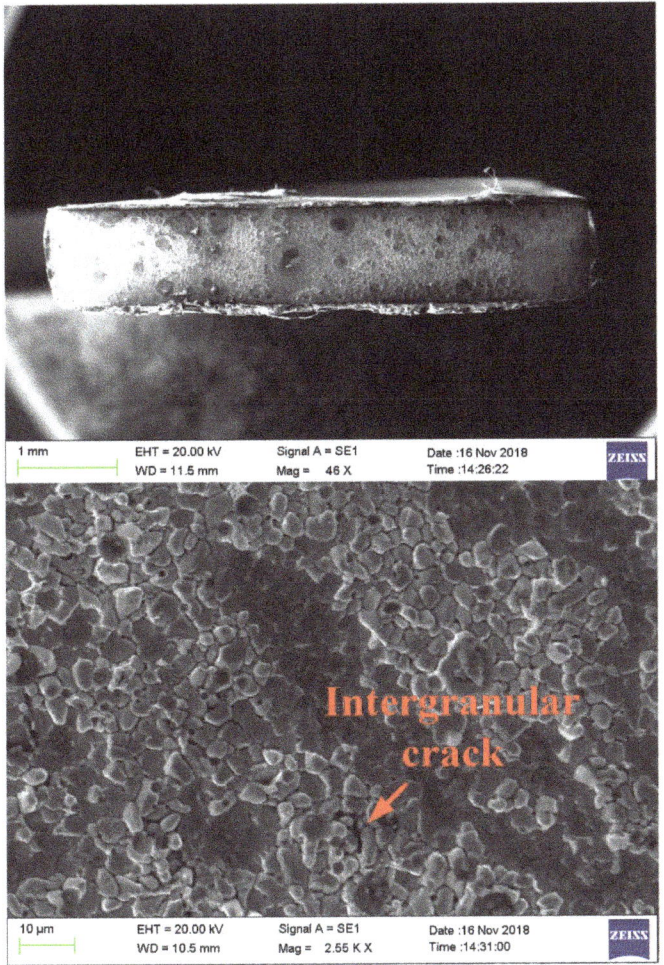

Figure 15. SEM graph of fracture surface at −80 °C (brittle).

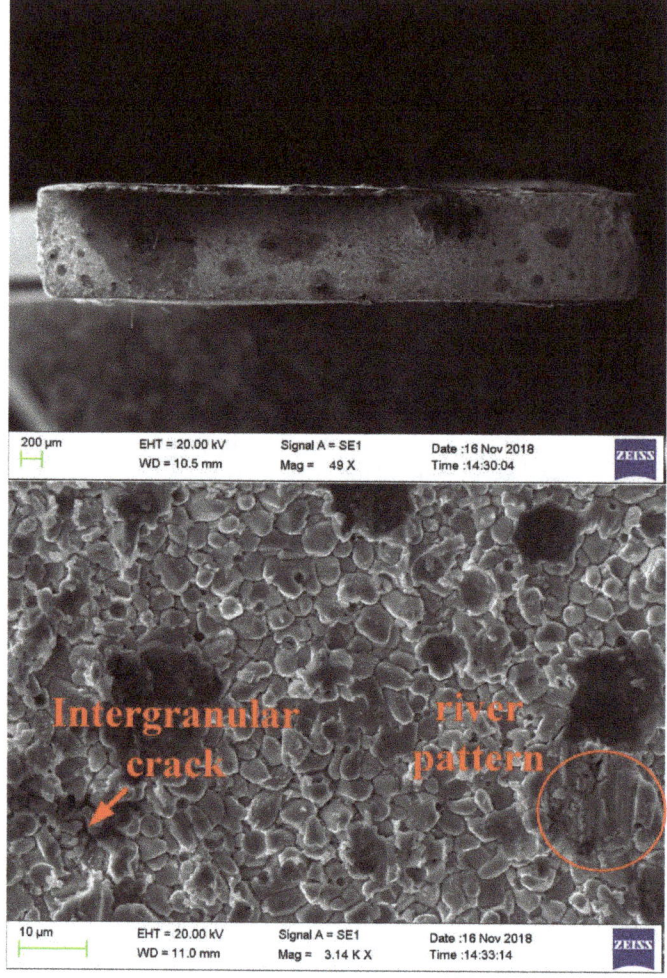

Figure 16. SEM graph of fracture surface at −100 °C (brittle).

5. Conclusions

In order to study the failure of a typical Pb-free circuit board, a series of methods was conducted to make failure analysis. Nondestructive examinations were used, and the failure position was located at the BGA solder joint. Then, destructive examinations were conducted, and the cause of failure was confirmed as the change of material characteristic under extremely low-temperature. A contrast test supported this result. Finally, verification experiments were conducted to verify the failure mechanism and find the transition temperature range of SAC305, which could make a guide for Pb-free circuits' deep space operation.

The results of failure analysis indicated the cause of failure. The Pb-free circuit board went through the shock tests at −100 °C. At this temperature, the fracture characteristic of SAC305 was brittleness, which was different from ductileness under room temperature. When the circuit underwent stress caused by shock test, it was easy for the solder joint to have cracks. When the crack grew to a specific length, the failure occurred. In the meanwhile, the transition temperature range of material property was also confirmed as −70—−80 °C.

Through this failure analysis case, it alerts us to focus attention on the reliability of Pb-free material applications for deep space exploration. Because of the low-temperature characteristics of SAC305, there may be a higher failure risk of actual operation. According to our study, the environment temperature needs to be kept higher than −70 °C. More studies need to be conducted to improve the reliability of Pb-free soldering used for aerospace components. Meanwhile, associated preventive methods, such as thermal preservation, are necessary.

Author Contributions: Conceptualization, B.W.; Methodology, Y.L.; Validation, Y.L.; Formal Analysis, G.F.; Investigation, Y.L.; Writing—Original Draft Preparation, Y.L.; Writing—Review and Editing, M.J.; Visualization, Y.L.; Supervision, W.Z. and X.Y. All authors have read and agreed to the published version of the manuscript.

Funding: This research received no external funding.

Acknowledgments: The authors are thankful to School of Reliability and Systems Engineering, Beihang University for providing the test equipment.

Conflicts of Interest: The authors declare no conflict of interest.

References

1. Wagner, S.; Zimmerman, D.; Wie, B. Preliminary Design of a Crewed Mission to Asteroid Apophis in 2029-2036. In Proceedings of the AIAA Guidance, Navigation, and Control Conference, Toronto, ON, Canada, 2–5 August 2010.
2. Chang, R. Influence of Microstructure and Cryogenic Temperature on Fatigue Failure of Indium Solder Joint. Ph.D. Thesis, University of Maryland, College Park, MD, USA, 2008.
3. Jullien, J.; Fremont, H.; Delétage, J. Conductive adhesive joint for extreme temperature applications. *Microelectron. Reliab.* **2013**, *53*, 1597–1601. [CrossRef]
4. Jiang, M.; Fu, G.; Wan, B.; Xue, P.; Qiu, Y.; Li, Y. Failure analysis of solder layer in power transistor. *Solder. Surf. Mt. Technol.* **2018**, *30*, 49–56. [CrossRef]
5. Jiang, N.; Zhang, L.; Liu, Z.-Q.; Sun, L.; Long, W.-M.; He, P.; Xiong, M.-Y.; Zhao, M. Reliability issues of lead-free solder joints in electronic devices. *Sci. Technol. Adv. Mater.* **2019**, *20*, 876–901. [CrossRef] [PubMed]
6. Directive EU. Restriction of the Use of Certain Hazardous Substances in Electrical and Electronic Equipment (RoHS). *Off. J. Eur. Communities* **2013**, *46*, 19–23.
7. Zhang, L.; Han, J.-G.; He, C.-W.; Guo, Y.-H. Reliability behavior of lead-free solder joints in electronic components. *J. Mater. Sci. Mater. Electron.* **2012**, *24*, 172–190. [CrossRef]
8. Jang, W.-L.; Wang, T.-S.; Lai, Y.-F.; Lin, K.-L.; Lai, Y.-S. The performance and fracture mechanism of solder joints under mechanical reliability test. *Microelectron. Reliab.* **2012**, *52*, 1428–1434. [CrossRef]
9. Ani, F.C.; Jalar, A.; Saad, A.A.; Khor, C.Y.; Ismail, R.; Bachok, Z.; Abas, M.A.; Othman, N.K. The influence of Fe2O3 nano-reinforced SAC lead-free solder in the ultra-fine electronics assembly. *Int. J. Adv. Manuf. Technol.* **2018**, *96*, 717–733. [CrossRef]
10. Kolenak, R.; Augustin, R.; Martinkovic, M.; Chachula, M. Comparison study of SAC405 and SAC405+0.1%Al lead free solders. *Solder. Surf. Mt. Technol.* **2013**, *25*, 175–183. [CrossRef]
11. Tsao, L.; Chang, S.-Y. Effects of Nano-TiO$_2$ additions on thermal analysis, microstructure and tensile properties of Sn3.5Ag0.25Cu solder. *Mater. Des.* **2010**, *31*, 990–993. [CrossRef]
12. Yao, Y.; Long, X.; Keer, L. A Review of Recent Research on the Mechanical Behavior of Lead-Free Solders. *Appl. Mech. Rev.* **2017**, *69*. [CrossRef]
13. Choudhury, S.F.; Ladani, L. Effect of Intermetallic Compounds on the Thermomechanical Fatigue Life of Three-Dimensional Integrated Circuit Package Microsolder Bumps: Finite Element Analysis and Study. *J. Electron. Packag.* **2015**, *137*. [CrossRef]
14. Mu, D.; McDonald, S.; Read, J.; Huang, H.; Nogita, K. Critical properties of Cu 6 Sn 5 in electronic devices: Recent progress and a review. *Curr. Opin. Solid State Mater. Sci.* **2016**, *20*, 55–76. [CrossRef]
15. Fazal, M.A.; Liyana, N.; Rubaiee, S.; Anas, A. A critical review on performance, microstructure and corrosion resistance of Pb-free solders. *Measurement* **2019**, *134*, 897–907. [CrossRef]

16. Pin, S.; Fremont, H.; Gracia, A. Lead free solder joints characterisation using single lap shear tests. In Proceedings of the 2017 18th International Conference on Thermal, Mechanical and Multi-Physics Simulation and Experiments in Microelectronics and Microsystems (EuroSimE), Dresden, Germany, 3–5 April 2017; pp. 1–6.
17. Gu, J.; Lin, J.; Lei, Y.; Fu, H. Experimental analysis of Sn-3.0Ag-0.5Cu solder joint board-level drop/vibration impact failure models after thermal/isothermal cycling. *Microelectron. Reliab.* **2018**, *80*, 29–36. [CrossRef]
18. Surendar, A.; Akhmetov, L.G.; Ilyashenko, L.K.; Maseleno, A.; Samavatian, V. Effect of Thermal Cycle Loadings on Mechanical Properties and Thermal Conductivity of a Porous Lead-Free Solder Joint. *IEEE Trans. Components Packag. Manuf. Technol.* **2018**, *8*, 1769–1776. [CrossRef]
19. Cai, Z.; Zhang, Y.; Suhling, J.C.; Lall, P.; Johnson, R.W.; Bozack, M.J. Reduction of lead free solder aging effects using doped SAC alloys. In Proceedings of the 2010 Proceedings 60th Electronic Components and Technology Conference (ECTC), Las Vegas, NV, USA, 1–4 June 2010; pp. 1493–1511.
20. Liu, F.; Meng, G. Random vibration reliability of BGA lead-free solder joint. *Microelectron. Reliab.* **2014**, *54*, 226–232. [CrossRef]
21. Xiaoguang, L.; Xiaoming, J.; Lichao, C.; Wengang, Z.; Xianfen, L.; Zhou, W.; Liu, X.; Jiang, X.; Cao, L.; Zhai, W.; et al. Microstructure and properties of Sn-3.8Ag-0.7Cu-xCe lead-free solders with liquid-liquid structure transition and Ce addition. *Mater. Res. Express* **2019**, *6*. [CrossRef]
22. Aamir, M.; Tolouei-Rad, M.; Din, I.U.; Giasin, K.; Vafadar, A. Performance of SAC305 and SAC305-0.4La lead free electronic solders at high temperature. *Solder. Surf. Mt. Technol.* **2019**, *31*, 250–260. [CrossRef]
23. Niu, X.; Zhang, Z.; Wang, G.; Shu, X. Low silver lead-free solder joint reliability of VFBGA packages under board level drop test at −45 °C. In Proceedings of the 2014 15th International Conference on Electronic Packaging Technology Conference, Chengdu, China, 12–15 August 2014; pp. 762–765. [CrossRef]
24. Liu, D.S.; Hsu, C.L.; Kuo, C.Y.; Huang, Y.L.; Lin, K.L.; Shen, G.S. Effect of low temperature on the micro-impact fracture behavior of Sn-Ag-Cu solder joint. In Proceedings of the 2010 5th International Microsystems Packaging Assembly and Circuits Technology Conference, Taipei, China, 20–22 October 2010; pp. 1–3.
25. Cheng, X.; Liu, C.; Silberschmidt, V. Numerical analysis of response of indium micro-joint to low-temperature cycling. In Proceedings of the 2010 11th International Conference on Electronic Packaging Technology & High Density Packaging, Beijing, China, 10–13 August 2009; pp. 290–293.
26. Liu, D.-S.; Hsü, C.-L.; Kuo, C.-Y.; Huang, Y.-L.; Lin, K.-L.; Shen, G.-S. A novel high speed impact testing method for evaluating the low temperature effects of eutectic and lead-free solder joints. *Solder. Surf. Mt. Technol.* **2012**, *24*, 22–29. [CrossRef]
27. Zachariasz, P.; Skwarek, A.; Illés, B.; Żukrowski, J.; Hurtony, T.; Witek, K. Mössbauer studies of β → α phase transition in Sn-rich solder alloys. *Microelectron. Reliab.* **2018**, *82*, 165–170. [CrossRef]
28. Lupinacci, A.; Shapiro, A.A.; Suh, J.O.; Minor, A.M. A study of solder alloy ductility for cryogenic applications. In Proceedings of the 2013 IEEE International Symposium on Advanced Packaging Materials, Irvine, CA, USA, 27 February–1 March 2013; pp. 82–88.
29. Lambrinou, K.; Maurissen, W.; Limaye, P.; Vandevelde, B.; Verlinden, B.; De Wolf, I. A Novel Mechanism of Embrittlement Affecting the Impact Reliability of Tin-Based Lead-Free Solder Joints. *J. Electron. Mater.* **2009**, *38*, 1881–1895. [CrossRef]
30. Ratchev, P.; Vandevelde, B.; Verlinden, B.; Allaert, B.; Werkhoven, D. Brittle to Ductile Fracture Transition in Bulk Pb-Free Solders. *IEEE Trans. Compon. Packag. Technol.* **2007**, *30*, 416–423. [CrossRef]
31. Xue, D. SN-Based Solder Joints Property and Life Prediction in Extremely Low Temperature. Master's Thesis, Harbin Institute of Technology, Harbin, China, 2015.
32. Ganesan, S.; Kim, G.; Wu, J.; Pecht, M.G.; Felba, J. Lead-free Assembly Defects in Plastic Ball Grid Array Packages. In Proceedings of the Polytronic 2005-5th International Conference on Polymers and Adhesives in Microelectronics and Photonics, Warsaw, Poland, 23–26 October 2006; pp. 219–223.

© 2020 by the authors. Licensee MDPI, Basel, Switzerland. This article is an open access article distributed under the terms and conditions of the Creative Commons Attribution (CC BY) license (http://creativecommons.org/licenses/by/4.0/).

Article

Reconstruction of Pressureless Sintered Micron Silver Joints and Simulation Analysis of Elasticity Degradation in Deep Space Environment

Wendi Guo [1], Guicui Fu [1], Bo Wan [1,*] and Ming Zhu [2]

[1] School of Reliability and Systems Engineering, Beihang University, Beijing 100191, China; gwd14141068@buaa.edu.cn (W.G.); fuguicui@buaa.edu.cn (G.F.)
[2] China Academy of Space Technology, Beijing 100094, China; baofan33@gmail.com
* Correspondence: wanbo@buaa.edu.cn; Tel.: +86-156-5292-8449

Received: 22 July 2020; Accepted: 9 September 2020; Published: 12 September 2020

Featured Application: It is a hot topic to find green adhesive materials to adapt to the deep space environment. Due to its economy, excellent electrical and thermal conductivity and mechanical properties, pressureless sintered micron silver paste has great application potential in the aerospace field. Several reliability issues with this material are mainly focused on its high temperature stability, while the microstructural evolution and macroscopic performance in the harsh deep space environment have not been considered. Moreover, the inevitable existence of pores caused by the specific sintering mechanism will significantly affect the performance of joints and result in potential reliability problems, and the relationship is not easily tested. Therefore, using a cost-effective method to study this relationship is necessary to promote its reliable applications. In this work, we design a test profile to stimulate the deep space environment, develop a simplified reconstruction and simulation methodology and quantitatively evaluate the elastic performance of joints. Furthermore, we also present the mechanism by which microstructural evolution has a negative impact on elastic mechanical performance in this environment.

Abstract: With excellent economy and properties, pressureless sintered micron silver has been regarded as an environmentally friendly interconnection material. In order to promote its reliable application in deep space exploration considering the porous microstructural evolution and its effect on macroscopic performance, simulation analysis based on the reconstruction of pressureless sintered micron silver joints was carried out. In this paper, the deep space environment was achieved by a test of 250 extreme thermal shocks of −170 °C~125 °C, and the microstructural evolution was observed by using SEM. Taking advantage of the morphology autocorrelation function, three-dimensional models of the random-distribution medium consistent with SEM images were reconstructed, and utilized in further Finite Element Analysis (FEA) of material effective elastic modulus through a transfer procedure. Compared with test results and two analytical models, the good consistency of the prediction results proves that the proposed method is reliable. Through analyzing the change in autocorrelation functions, the microstructural evolution with increasing shocks was quantitively characterized. Mechanical response characteristics in FEA were discussed. Moreover, the elasticity degradation was noticed and the mechanism in this special environment was clarified.

Keywords: pressureless sintered micron silver joints; deep space environment; extreme thermal shocks; reconstruction; simulation; elastic mechanical properties

1. Introduction

In a deep space environment, the exploration equipment with complex electric systems suffers from extreme thermal shocks, inducing material performance degradation and further leading to the failure of electronic packaging. The reasons could be clarified by the research on the reliability of Sn/Pb and SnAgCu solder packaging which shows vulnerability to thermal shocks of Pb-free solders and Sn/Pb solders [1,2]. Therefore, finding alternative green bonding materials to adapt to the space environment has become an urgent research hotspot. Considering its outstanding ability to withstand heat, power, and stable mechanical properties [3–5], the sintered silver material has broad potential applications in harsh environments. However, compared to nano-silver particles [6–8], the pressureless sintered micron silver is affordable but is given less attention.

To reliably put it into use in deep space, the study of mechanical properties and the possible degradation mechanism of pressureless sintered micron silver joints in a deep space environment is valuable. Due to specific sintering mechanisms, randomly distributed pores inevitably exist in the microstructure of sintered joints [9], which will significantly affect material properties (i.e., the mechanical, thermal properties, etc.) and lead to possible reliability problems. The published studies drew the conclusion that Spherical and cylindrical voids had a significant effect on the thermal resistance of CSP packages [10], and an increase in void rate could result in a decrease in the shear strength of the solder layer [11], and voids would greatly shorten the fatigue life due to reduction in the overall carrying capacity [12]; these studies all established regular pore models and applied different pore locations and distribution rates. This kind of simulation study is mainly used to analyze the relationship between the microstructure and macroscopic performance of the porous adhesive layer because of costly and technically demanding experiments. However, the pores were simplified to the circle or column shape regardless of the actual shape in these simulations, so the corresponding results could be less accurate. For obtaining precise results, other researchers worked to link real microscopic structures of sintered silver with the properties. T. Youssef [13] reconstructed the 3D model of a sintered sample by utilizing a focused ion beam–scanning electron microscope (FIB–SEM) and the software AVIZO, and analyzed the changing trend of thermal and mechanical properties with increased porosity, which required a large number of high-accuracy serial slice images. X. Milhet [14] obtained elastic constants of sintered joints by applying dynamic resonant testing to sintered bulk specimens which were produced to represent the real structure, but this needed large expenditure.

In order to reconstruct the microstructure of sintered silver and predict its properties in an economical, precise and practical way, the correlation function method based on the probability and mathematical statistics theories is introduced. Correlation functions have been developed to describe random heterogeneous materials, including n-point correlation functions, surface correlation functions, the linear path function, chord-length density function and so on [15]. Among them, n-point correlation functions can take the shape, distribution, and orientation of material components into consideration, and have shown themselves to be feasible in the numerical simulation of isotropic and anisotropic media, where n represents the order of functions. With a higher order, n-point correlation functions could provide more precise characterization of heterogeneous microstructures [16–18], but the technique of obtaining the optimum approximation with effective, unbiased, and accurate experimental estimation from projected images is not yet mature. The second order correlation function (two points) has successfully reconstructed porous media such as concrete [19], Berea sandstone [20] and other composites [21,22], whilst retaining the microstructural features. However, few simulations studying the microstructure of micron silver sintered joints which have similar porous morphology to the above materials and their relationship with mechanical properties based on this method have been performed, as a result of complex calculations and lacking of an approach to transfer the reconstructed models into Finite Element Analysis models.

To solve the above problems, a thermal shock test of −170 °C~125 °C is conducted to simulate the deep space environment, and the section morphology are obtained for further study. A simplified reconstruction method of pressureless sintered micron silver joints is presented and used in simulation analysis of elasticity degradation, which is a key parameter to evaluate mechanical properties in the deep space environment. The rest of this paper is organized as follows: Section 2 describes the designed thermal shock test and sample information simulating the real package structure, as well as SEM image acquisition required for modeling. Section 3 presents detailed methods of the morphology characterization and reconstruction of joints and proposed simulation procedure, and then verifies these methods with relative entropy, analytical models and test results. Section 4 analyzes the microstructure evolution during extreme thermal shocks, and the mechanical response characteristics, and discusses the negative effect of microstructure on elastic properties of joints. Finally, in Section 5, the concluding remarks are stated.

2. Experiments

2.1. Die Attachment Samples

As shown in Figure 1a, the polished side of a square silicon die (5 mm × 5 mm × 1 mm) was coated with a 50 nm Ti and a 50 nm Ag metallization layer using magnetron sputtering technology. A Ti layer was used as an adhesive by reacting it with natural oxides on the wafers, and an Ag layer provided a covering layer for tight integration with sintered micron silver paste. For the die bonding substrate (10 mm × 10 mm × 2 mm), copper (Cu 99.9%) was selected to make it, and a 50 nm Ti layer designed to prevent oxidation and copper atoms from spreading to sintered Ag joints was sputtered on one side, followed by a 2 μm Ag layer. The sandwich structure of the die attachment sample is shown in Figure 1b, where the die attachment structure was the micron silver paste with a thickness of 100 μm. The sample was assembled by brushing the paste on the metallized substrate, placing the silicon wafer on the Ag paste with tweezers, and sintering in air at the temperature of 230 °C without pressure, as shown in Figure 1c.

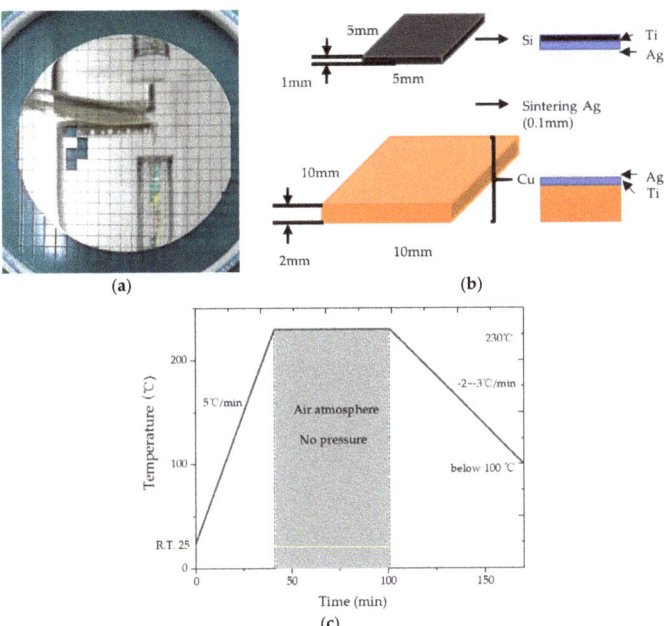

Figure 1. Details of die attachment samples: (**a**) Ti/Ag plated silicon dies; (**b**) the die attachment structure consisting of a Cu substrate, micron silver paste and one side polished die; (**c**) the curing process of micron silver joints.

Material parameters of the micron silver paste are listed in Table 1, and the SEM image with spectrum spot and EDS analysis result are shown in Figure 2. It can be seen that the material consists of submicron silver particles and silver flakes and involves the negligible carbon content which shows a small peak.

Table 1. Parameters of the pressureless sintered micron silver material.

Parameter	Condition	Micron Ag Paste
Ag content (wt%)	in paste	92–94
Viscosity, E-type 3°cone (Pa·s)	5 rpm@RT	18–22
Thixotropic Index	0.5 rpm/5 rpm	6.5
Volume Resistivity (Ω·cm)		5×10^{-6}
Poisson's ratio		0.25
Density (g/cm^3)		7.8
Elastic modulus (GPa)	Nanoindentation@RT	25
CTE (ppm/°C)	Isotropy	18

 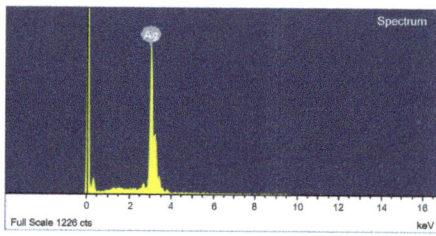

Figure 2. SEM image with spectrum spot and EDS analysis.

2.2. Thermal Shock Test

The drastic change of ambient temperature in deep space missions is an important factor resulting in internal defects in the packing of electronic devices. In order to study the mechanical properties of this new bonding material in space missions, a thermal shock test was carried out to simulate the aerospace environment, in which the temperature changed from −170 °C to 125 °C covering the temperature range of the moon, Mars, common asteroids and comets. During the test, samples were placed in a high and low temperature chamber with a thermal shock profile (Figure 3). The soak time of extreme temperature was 15 min, and the frequency was about 30 min/cycle.

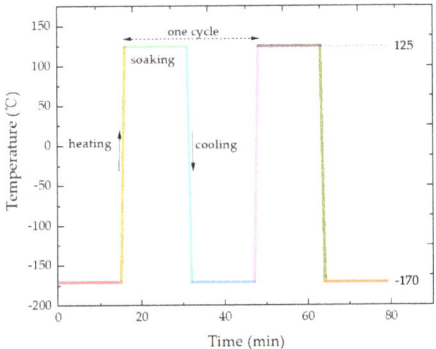

Figure 3. Thermal shock profile for the test application.

2.3. SEM Images of Micron Silver Sintered Joints

Prior to the thermal shock test, the die shear strength test was conducted to evaluate bonding strength and the reliability of bonding at the interfaces. The average shear strength of the sintered micron silver samples was 15 ± 2 Mpa, which showed good bonding quality. During the shearing process, a fracture almost occurred through the adhesive layer, indicating the reliability of bonding at the interfaces. However, there was still delamination occurring at the interface between the micron silver paste and the substrate, as seen in Figure 4. This was related to defects in the metallized layer. Samples that failed in such way in the further thermal shock test needed to be removed to focus on the adhesive layer performance degradation.

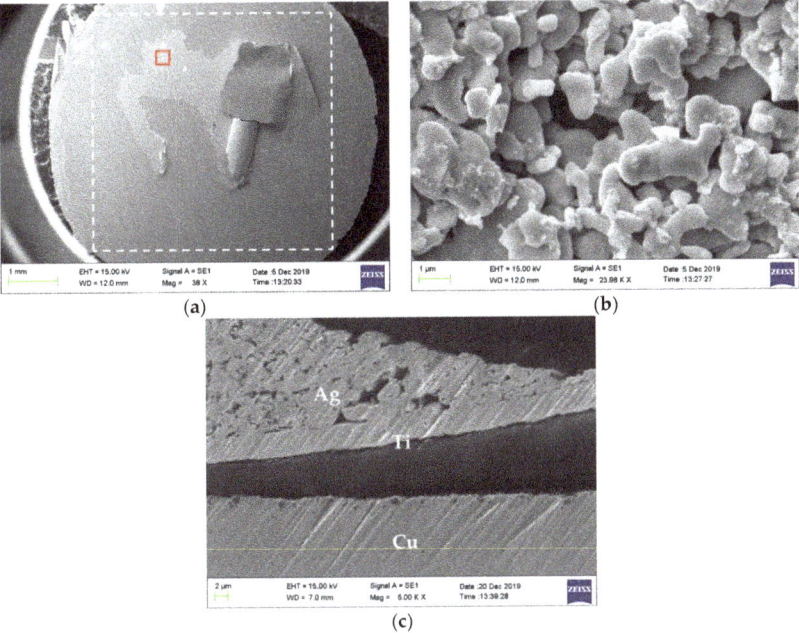

Figure 4. SEM images of the fracture of a micron silver joint after shear testing. The white frame represents the die position: (**a**) fracture occurs at the substrate–joint interface; (**b**) larger view of micron silver particles coalescing from selected red region in (**a**); (**c**) the metallized layer is separated from the substrate.

Sintered silver joints are characterized by a typical porous structure. Samples were taken out and molded every few thermal shocks. After longitudinal grinding, two-dimensional (2D) images of micron silver sintered joints were obtained by observing the microscopic morphology and shown in Figure 5, which were used as the initial data for three-dimensional (3D) reconstruction. No significant cracks were found in destructive tests.

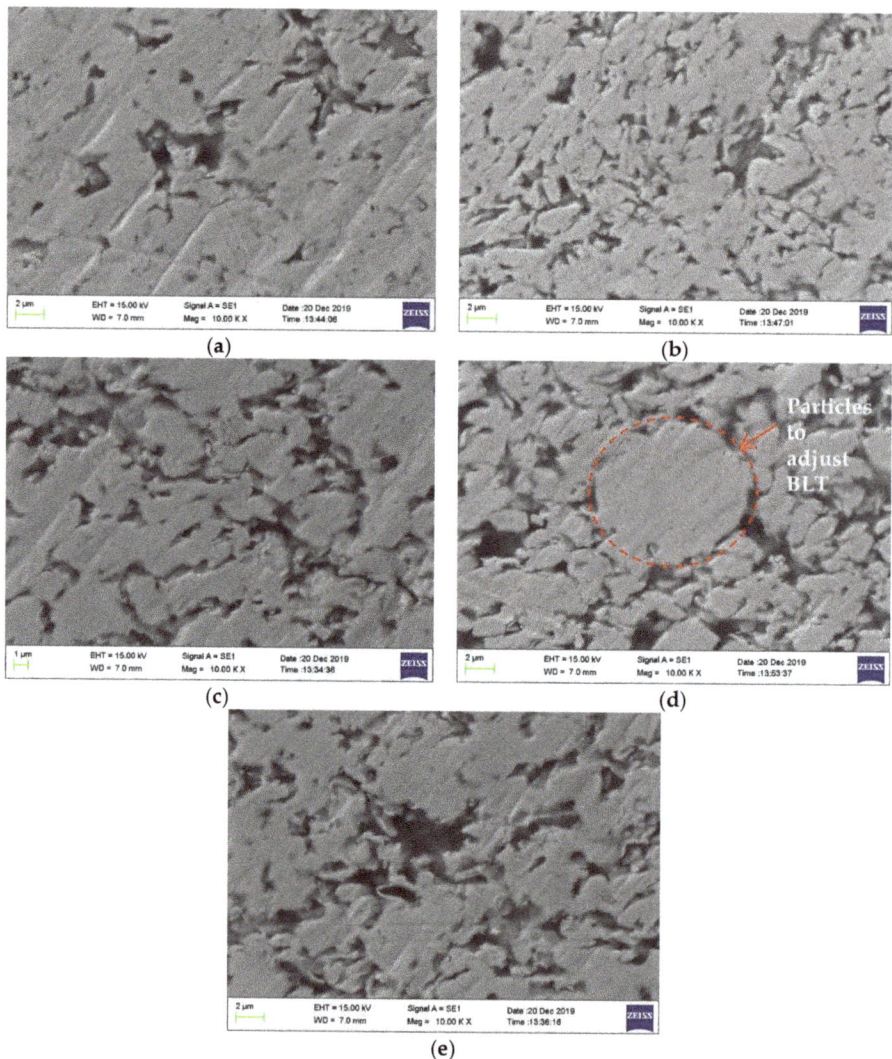

Figure 5. SEM images of micron silver sintered joints under thermal shocks for: (**a**) 0; (**b**) 50; (**c**) 100; (**d**) 150; and (**e**) 250 cycles.

3. Model Reconstruction and Finite Element Analysis

3.1. Structural Characterization and Reconstruction (SCR)

The Joshi Quiblier Adler (JQA) method [23] is a morphology autocorrelation-function-based tool for reconstructing porous media. To study the relationship between microstructure and performance of micron silver sintered joints from a microscopic perspective, the method was simplified and used to characterize 2D cross sections and reconstruct a two-phase heterogeneous 3D model to provide the high-dimensional point cloud data required for simulation. During reconstruction, a Gaussian random field was used to generate spatial media, and the morphology and dispersion of two-phase interface were described by autocorrelation function. A region-based image segmentation technique iteratively solved the porosity of 3D models which was consistent with original images, and these models were corrected by smoothing operation. The whole flow is illustrated in Figure 6.

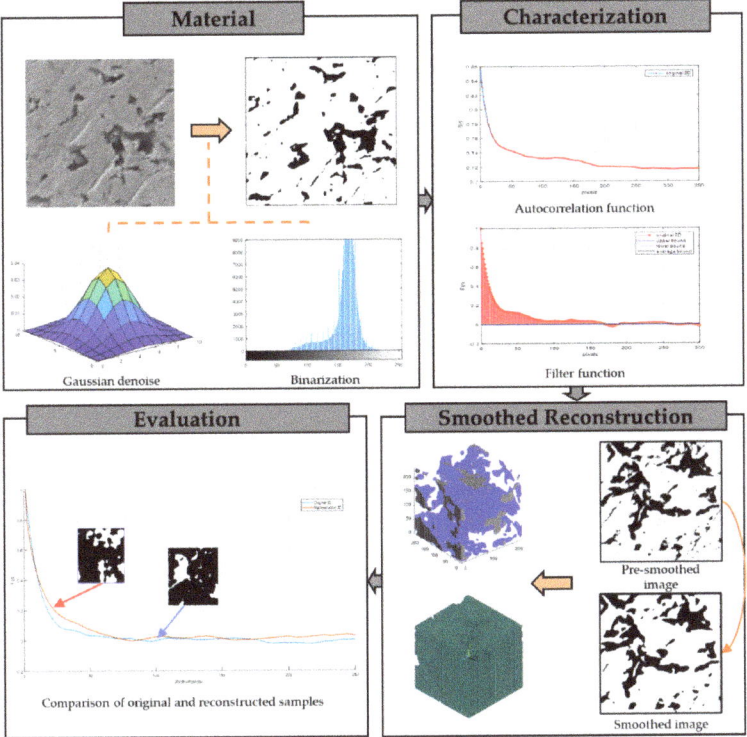

Figure 6. The Flow chart of reconstructing stochastically equivalent 3D morphology of sintered micron silver joints.

There are four main steps required to reconstruct the randomly distributed medium from an SEM picture using SCR, which are given below:

1. Denoising, threshold segmentation and binarization processing are applied on the original image to get the two-phase random medium, which is expressed as $V(\omega) \in \mathbb{R}^3$, a spatial domain. Where ω is the domain intercepted from the probability space of volume V, including two parts: the pore volume fraction φ_1 in the region V_1 and the volume fraction of micron silver particles φ_2 in the region V_2. Binary porous media may be represented by an indicator function $I(x)$, as defined below:

$$I(x) = \begin{cases} 1, x \in V_1 \\ 0, x \in V_2 \end{cases}. \tag{1}$$

2. The two-point autocorrelation function S2(r) is used to describe the morphology as shown in Equations (2) and (3). S2(r) is defined as follows: two arbitrary points, x1 and x2 of the distance r,

are selected in an observation region, and the probability that both points are in one phase is S2(r), which is illustrated in Equation (4). For the isotropic material in this study, it can be calculated by bilinear interpolation [24].

$$S_2^i(r) = \langle I^i(x_1) I^i(x_2) \rangle = P\{I^i(x_1) = 1, I^i(x_2) = 1\}, \tag{2}$$

$$\begin{cases} S_2^i(r) = \phi_i, r = 0 \\ \lim_{r \to \infty} S_2^i(r) = \phi_i^2 \end{cases}, \tag{3}$$

$$S_2(r) = \frac{\sum_{(m,n) \in \Omega} \left[\sum_{i=1}^{M} \sum_{j=1}^{N} I_{i,j} I_{i+m,j+n} \right]}{\omega MN}, \tag{4}$$

where ω is the number of elements in the set Ω which is calculated in Equation (5):

$$\Omega = \{(m,n) | m^2 + n^2 = r^2, r \leq [\min\{M,N\}/2] \}. \tag{5}$$

The normalized autocorrelation function in the spherical coordinate, as Equation (6), is used as the filtering function of normally distributed noise:

$$F(R) = \frac{E\{[I^i(x+R) - \phi_i] \cdot [I^i(x) - \phi_i]\}}{E\{[I^i(x) - \phi_i]^2\}}. \tag{6}$$

3. X1 and X2 are uniformly distributed random numbers. Based on Box–Muller, these two random numbers can be used to generate Gaussian-distributed noise N efficiently, with a mean of 0 and a variance of 1, as shown in Equation (7).

$$N = \sqrt{-2 \ln X_1} \cos(2\pi X_2), X_1 \sim U(0,1), X_2 \sim U(0,1). \tag{7}$$

As shown in Equation (8), the initial 3D image with Gaussian noise can be obtained:

$$I(i,j,k) = \sum_{r,s,t} N(r,s,t) \times F(i+r, j+s, k+t). \tag{8}$$

According to Equation (9), iterative threshold segmentation is performed on 3D images to match the porosity of SEM pictures. The Fourier transform is used to perform three-dimensional Gaussian smoothing operation to correct reconstructed 3D models, which is calculated as Equation (10). Finally, high-dimensional binary matrix is then obtained:

$$porosity(\%) = \frac{\sum_{\substack{1 \leq i \leq M \\ 1 \leq j \leq N}} I_{i,j}^{(V_1)}}{MN} \times 100\%, \tag{9}$$

$$f(x,y,z) = 2\pi^{-\frac{3}{2\sigma^3}} \exp\left(-\left(\frac{2x^2}{\sigma^2} - \frac{2y^2}{\sigma^2} - \frac{2z^2}{\sigma^2}\right)\right). \tag{10}$$

4. Kullback–Leibler (KL) divergence, also known as relative entropy, is a measure of the difference between two distributions P1 and P2 to evaluate the reconstruction quality, as shown in

Equation (11). The KL divergence is calculated from 0 to $+\infty$, indicating the similarity from the most to the least.

$$KL(P_1\|P_2) = \sum_{x \in X} P_1(x) \log \frac{P_1(x)}{P_2(x)}. \quad (11)$$

Through the above process, a series of high-dimensional data equaling to the 3D geometry reconstructed model could be obtained and transferred into the Finite Element Analysis (FEA) model as below. The high-dimensional data can be discretized and reduced into a set of unit information, called volume data. These data logically form a 3D array space, and each array point stores volume location and feature information, called a voxel. The location of one voxel was determined by layer, row and column, as shown in Figure 7a. Voxels belonging to the medium are marked as v = 1, and those belonging to the pore are marked as v = 0. The thin layer of volume data is shown in Figure 7b. All voxels' information is stored in a TXT file and imported in the ANSYS Parametric Design Language (APDL) program to build the 3D entity in ANSYS Mechanical for FEA, which is a Boolean description of micron silver sintered material with voxels as units in space. The voxel is built by 8 key-points illustrated in Figure 7c, where i, j and k represent the voxel location coordinate, respectively.

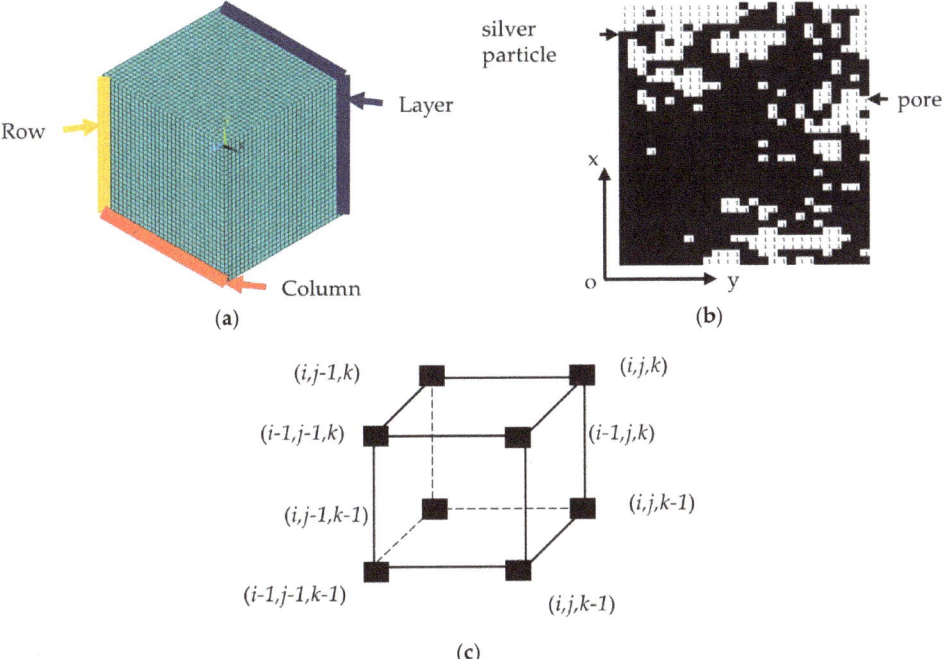

Figure 7. Indication of spatial data. (**a**) Spatial volume data; (**b**) A thin-layer model; (**c**) One voxel built in ANSYS Mechanical.

Without considering the effect of the grain boundary of fused micron silver particles, reconstructed models of joints can be obtained by reverse filling the pores with Boolean operation. Reconstructed pore visualization results and 3D FEA models corresponding to SEM images of micron silver sintered joints under different thermal shocks are shown in Figure 8.

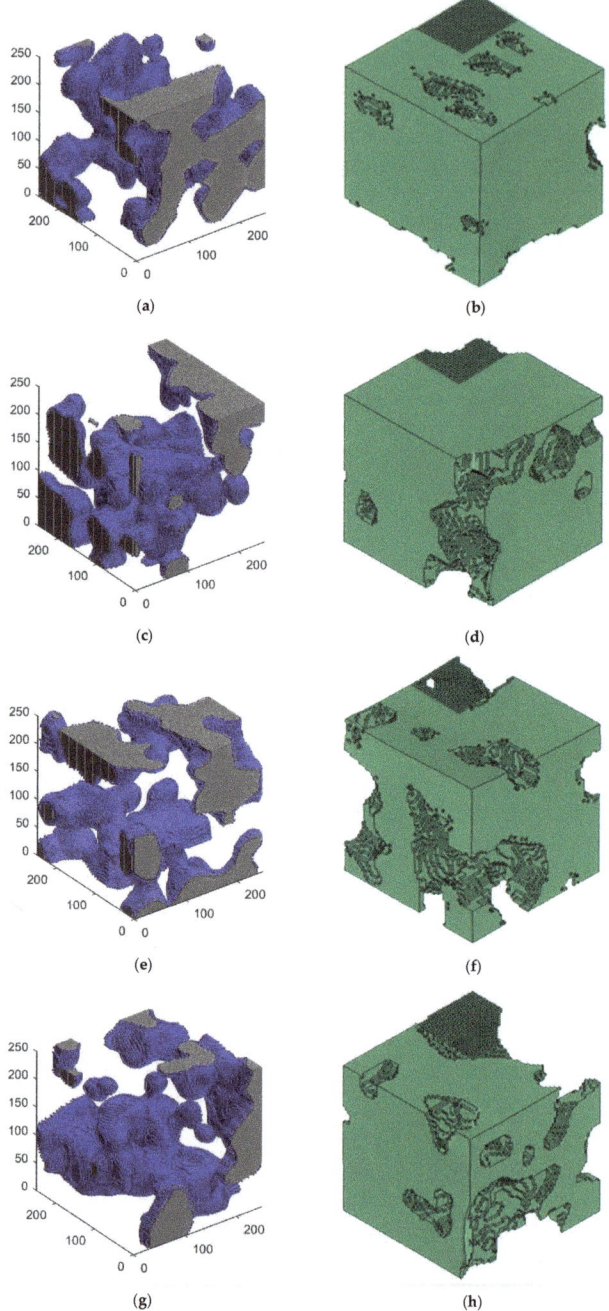

Figure 8. Reconstructed pore topological structures and 3D Finite Element Analysis (FEA) models with mesh in a small area corresponding to SEM images of micron silver sintered joints under different thermal shocks: (**a**,**b**) 50; (**c**,**d**) 150; (**e**,**f**) 150; (**g**,**h**) 250 cycles. In 3D views of the left column, gray represents defined geometric boundaries, and blue is the iso-surface.

Normalized autocorrelation functions of SEM images and 3D reconstruction models are plotted in Figure 9, where the size of reconstructed models (250 pixels) is much larger than the observed correlation length of SEM samples (the autocorrelation function asymptotic location), illustrating reconstructed models can represent the microstructure of sintered joints. The KL divergence values are calculated and listed in Table 2, which are within an acceptable range (less than 15%). Both qualitative observation and quantitative calculation show that the reconstructed models are consistent with the original images.

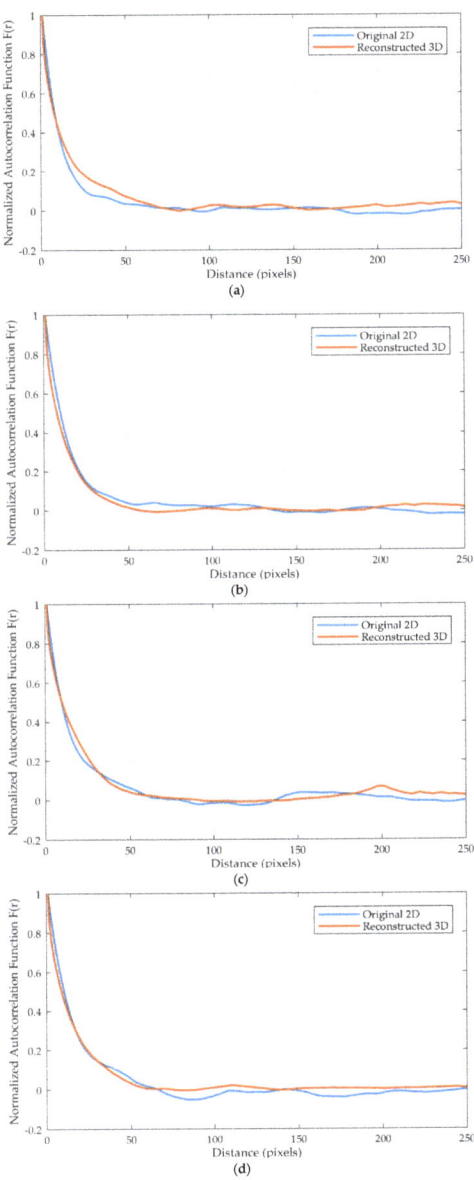

Figure 9. Comparison of autocorrelation functions of SEM images and 3D reconstruction models with thermal shock cycles are (**a**) 50; (**b**) 100; (**c**) 150; (**d**) 250.

Table 2. Kullback–Leibler (KL) divergence between autocorrelation functions.

Thermal Shock (Cycles)	50	100	150	250
KL divergence	0.11	0.12	0.12	0.08

3.2. FEA Simulation

To obtain the effective elastic modulus value of the micron silver sintered joints under thermal shock, the simulation loading conditions were set as below: one side of the reconstructed model was subjected to a fixed constraint and the opposite side was put into stress σ_i, shown in Figure 10. The input Young's modulus was set from nanoindentation result of 25 Gpa. The density and the Poisson's ratio were 7.8 g/cm^3 and 0.25. For the FEA model with length L, the displacement on the force surface and equivalent stress of each node are extracted, and the effective elastic modulus is calculated by Equations (12) and (13).

$$E_i = \frac{F_i L}{A x_i}, \tag{12}$$

$$E = \frac{\sum_i E_i}{n} = \frac{\sum_i \left[\sigma_{ave} / \left(\frac{dL_i}{L}\right)\right]}{n}, \tag{13}$$

where E_i is the effective elastic modulus obtained by fixing X, Y and Z planes, respectively, to eliminate the calculation error caused by structural randomness. Fi is tension, x_i is displacement and A is the section area. n is the number of simulation tests, and σ_{ave} is the average tensile stress which is calculated as Equation (14). dL_i is the average displacement in the stress direction.

$$\sigma_{ave} = \frac{\sum_i \sigma_i \times V_i}{V_{medium}}, \tag{14}$$

σ_i is the equivalent stress in the ith node. V_i is the volume of element in which node i resides. V_{medium} is the total volume of all nodes in elements.

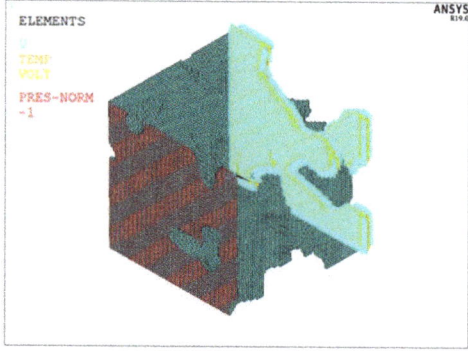

Figure 10. Simulation loading diagram. Green represents a fixed constraint, and red represents applied tensile stress. The element type is Solid 226.

The calculation of average stress can reduce the stress inequality caused by complex structure, which is equivalent to the effective stress of sintered joints. Moreover, it depends on the microstructural characteristics rather than stress concentrations of individual points.

Table 3 shows the model size, the number of elements and calculation time during simulation. After generating FE models, the solving process generally requires 20 to 30 min. The cumulative time spent on modeling and meshing is relatively significant, about 3 h, so as to ensure the quality of these simulation models.

Table 3. Information about elastic modulus simulation.

Thermal Shock (Cycles)	Model Size (Pixels)	Number of Elements	Calculation Time (min)
50		172,195	
100	250^3	168,307	20 to 30
150		160,390	
250		160,380	

The simulation results of displacement and equivalent stress are shown in Figures 11 and 12.

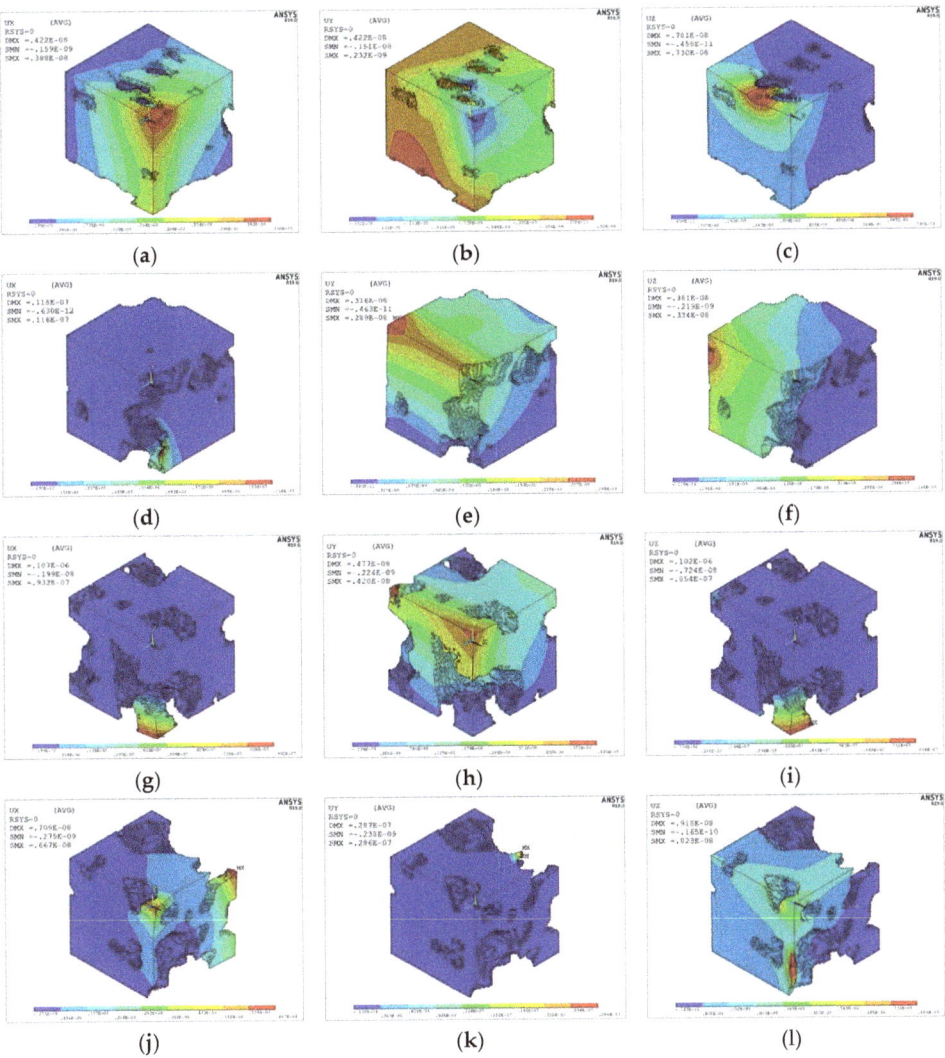

Figure 11. Displacement analysis results of micron silver sintered joints under different thermal shocks by fixing X, Y and Z planes, respectively: (a–c) 50; (d–f) 100; (g–i) 150; (j–l) 250 cycles.

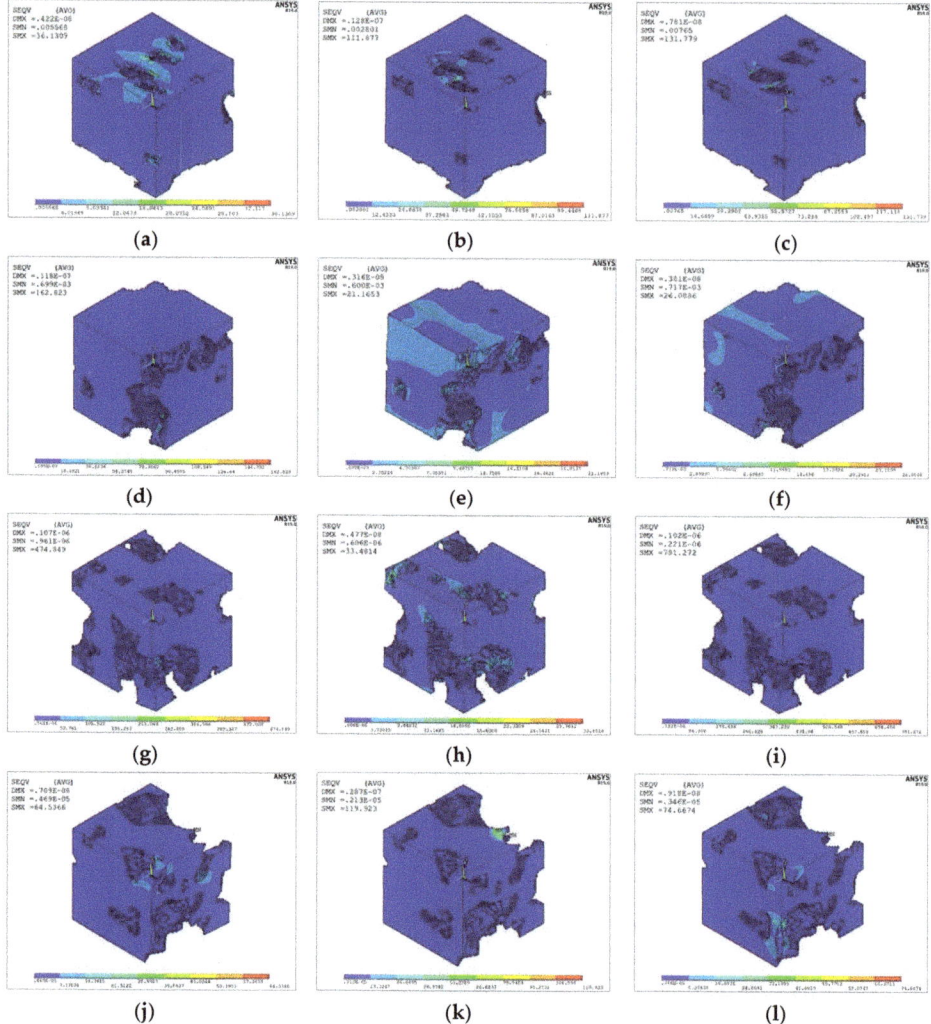

Figure 12. Equivalent stress analysis results of micron silver sintered joints under different thermal shocks by fixing X, Y and Z planes, respectively: (**a**–**c**) 50; (**d**–**f**) 100; (**g**–**i**) 150; (**j**–**l**) 250 cycles.

3.3. Analytical Model and Experimental Validation

The analytical model of the effective elastic constants of porous solids is based on numerical simulation and microstructure measurement, and it is considered to be a relatively accurate empirical equation for predicting elastic properties. In order to ensure the rationality of this simulation route and accurately quantify the microstructural damage on elastic properties of micron silver sintered joints, two analytical models were used to calculate the elastic modulus at different thermal shocks and these results were compared with simulation results. The first model corresponds to the Ramakrishnan and Arunachalam (R&A) method [25] and is conducted as Equations (15) and (16).

$$E_p = E_0(1-p)^2/(1+b_p p), \tag{15}$$

$$v = \frac{1}{4}\frac{(4v_0 + 3p - 7v_0p)}{(1 + 2p - 3v_0p)}, \tag{16}$$

where, b_p = 2 to 3v, p is the pore volume fraction. v_0, E_0 is the Poisson's ratio and elastic modulus for undamaged material, and E_p is the elastic modulus corresponding to a certain damage state.

Another model, the modified value of porous materials (M) [26] in Equation (17) was used to estimate the effect of microscopic damage on elastic modulus. The parameter definition is the same as above.

$$E_p = E_0 - pE_0 \left(\frac{9 - 4v_0 - 5v_0^2}{7 - 5v_0} \right). \tag{17}$$

Furthermore, nanoindentation test was used to measure the load and corresponding displacement of pressureless micron sintered joints. The geometric vertices and center point of the silver welding layer were selected as test positions, and the average value was taken as the test result. If there is a large deviation in the above positions, it is determined that the silver bonding layer of the sample is uneven during sintering, and needs to be rejected. Meanwhile, a new sample is re-selected for testing. The designed stress curve is shown in Figure 13.

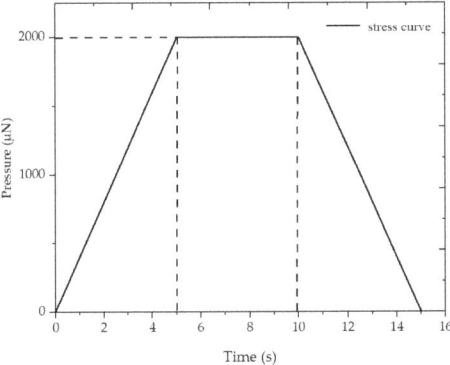

Figure 13. Applied nanoindentation stress curve.

For geometries based on SCR method, normalized calculation results of FEA and theoretical models and nanoindentation tests are plotted in Figure 14. The error among simulation and test and R&A method is within 15%, which indicates that the proposed simulation route and the elastic properties of micron silver sintered joints obtained by FEA are reliable and precise.

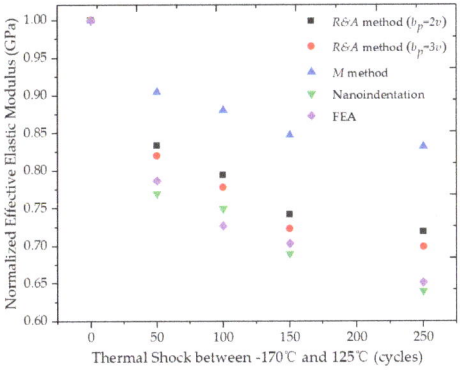

Figure 14. Relative comparison of FEA, analytical models and nanoindentation results.

4. Results and Discussion

4.1. Microstructural Evolution

The simulation illustrates that the effective elastic modulus degrades with the accumulation of shock cycles. In order to study the cause of performance degradation, the microstructural evolution is characterized from autocorrelation functions.

The specific surface, marked as s, of two-phase media is defined as the two-phase interface area per unit total volume, which can be obtained from the slope of the two-point autocorrelation function [27]. In this study, s of original images and reconstructed models is conducted from Equation (18):

$$\frac{d}{dr}S_2(r)\bigg|_{r=0} = \begin{cases} -s/\pi, D=2 \\ -s/4, D=3 \end{cases}. \tag{18}$$

Specific surface can be used to estimate the hydraulic diameter (D_H) of porous media [28], which is calculated in Equation (19). V is the volume fraction of the medium:

$$D_H = \frac{4V}{s} \tag{19}$$

The captured autocorrelation functions of original SEM images from Figure 5 changes with thermal shocks are shown in Figure 15. There is an increase in the slope at the origin and a decrease in the correlation length. Changes of the mean particle size, hydraulic diameter and specific surface in SEM pictures and reconstructed models are calculated as shown in Table 4.

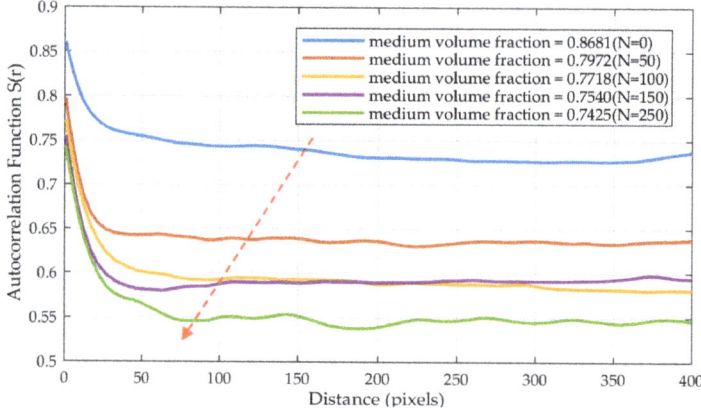

Figure 15. Autocorrelation functions under the accumulation of thermal shocks.

Table 4. The particle size, hydraulic diameter and specific surface of original images and reconstructed models.

Thermal Shock (Cycles)	$D_{P,2D}$ (µm)	$D_{P,3D}$ (µm)	$D_{H,2D}$ (µm)	$D_{H,3D}$ (µm)	s_{2D} (µm^{-1})	s_{3D} (µm^{-1})
0	4.30	-	3.16	-	1.10	-
50	3.03	2.88	1.88	2.03	1.70	1.57
100	2.24	2.12	2.25	2.16	1.37	1.43
150	1.91	1.82	2.15	2.05	1.40	1.47
250	2.03	1.97	1.98	1.86	1.50	1.60

It can be seen that the average micron-silver particle size decreases, while this is in contrast to a slight increase in the size of silver grains (−45 °C~250 °C) reported in [29]. It should be noted that the surface diffusivity of silver is relatively lower at 125 °C [30] than the melting point and sintering temperature. Meanwhile the soak of −170 °C seems to prevent particle recrystallization growth and even reverse-driven particles from dispersing. The hydraulic diameter of the medium declines with fluctuation. If the porous medium is regarded as an interconnected pipeline network, the above situation can be interpreted as the rise of pore density and pore nucleation growth induced by cyclic loading, which is consistent with the measured increase in average pore size and porosity [31]. The whole tendency of the specific surface goes up, which physically means that the total surface area of pores increases. The results demonstrate that damage gradually accumulates with thermal shock testing, resulting in the increase in porosity and the role of pores as grain growth inhibitors. The contact area between adjacent grains and the size of the medium decreases. The medium particles tend to disperse. In Figure 15 autocorrelation functions slightly oscillate after 50 thermal shocks, which reflects in the influence of the roundness of micron silver particles. Furthermore, another important finding from visual the view of pores (Figure 8) is that the pore distribution area becomes uneven and the local density changes, which will further promote the initiation of cracks.

4.2. Mechanical Response Characteristics

Extrema of mechanical response characteristics of nodal solutions in previous FEA simulation could be observed in Figures 11 and 12, which illustrate the displacement and stress of reconstructed micron silver sintered joints suffering different thermal shock cycles in the X, Y, and Z directions. It can be seen that the maximum and minimum values of the displacement and stress show a similar fluctuation in the X and Z directions with increasing thermal shocks. This is because the x–z plane represents the parallel interface of this sandwich structure, and the Y axis is the longitudinal constrained sintering direction, so there is a certain degree of anisotropy.

It can be found that the maximum displacement point appears near the junction of silver phase and larger pores, or the corner of pores, and the value shows an increasing trend, which demonstrates that the rise of porosity and particle dispersion caused by increased thermal shocks leads to the decrease in the size of the load-transmission area between particles, and can further result in a large displacement field across the local bearing surface. This situation also indicates that the elastic performance of micron silver sintered joints declines. On the other hand, the variation range of triaxial stress gradually increases as the thermal shock cycles increases. In addition, stress concentrations exist in all three directions. This situation could be explained by more likely stress concentrations resulting from the growing porosity.

Table 5 illustrates the triaxial elastic modulus and the effective elastic modulus by FEA, and the apparent density of samples. It can be seen that the Y-axis elastic modulus is a little greater than those of the X and Z axes. This phenomenon should be derived from the sample structure, and the constrained sintering state leads to a slight anisotropic microstructure [32]. Considering the slight degree of anisotropy, the effective elastic modulus is adopted to evaluate the degradation behavior of the micron sintered silver joints. As the thermal shock cycles increase, the apparent density of micron silver joints decreases, indicating that cyclic thermal stress can drop the material densification.

Table 5. The elastic modulus in three directions, the effective elastic modulus and apparent density.

Thermal Shock (Cycles)	E_X (GPa)	E_Y (GPa)	E_Z (GPa)	E_{eff} (GPa)	Apparent Density (g/cm^3)
50	19.63	20.70	19.50	19.94	7.46
100	17.60	20.98	19.08	19.22	7.22
150	16.31	18.12	17.52	17.32	7.05
250	16.22	16.92	15.56	16.24	6.93

4.3. Elasticity Degradation

Based on the data in Table 5, Figure 16 illustrates the change of effective elastic modulus and porosity of micron silver sintered joints with accumulated extreme thermal shocks. The porosity changes from 13.19% to 25.75% and the effective elastic modulus decreases by about 36%. Since the elastic modulus of three axes is calculated with deviation among different directions, it can be concluded that the elastic properties of micron silver joints are also related to the microscopic topological structure. The closer the reconstruction is to the original, the less the simulation fluctuates.

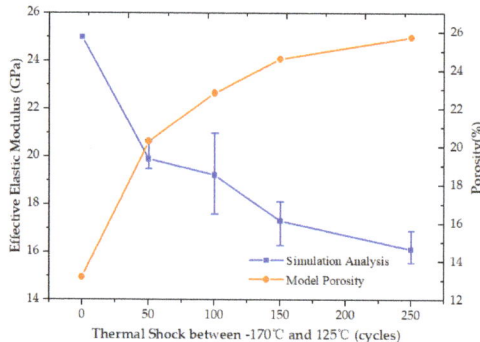

Figure 16. Changes in effective elastic modulus and porosity.

Compared with analytical predictions which only consider porosity as the main factor, it is found that FEA is close to the R&A method for solving the problem of modulus uncertainty by using the special model of the change of the effective Poisson's ratio. However, the forecast is slightly larger. The large deviation of the M method may be related to the premise that the composite material consists of a continuous matrix phase with a high concentration of rigid spherical inclusion suspension. The comparison with the test results also indicates that FEA can easily realize the prediction of elastic properties in micron silver sintered joints. The calculation taking errors caused by structural randomness into account basically covers nanoindentation results (Figure 17).

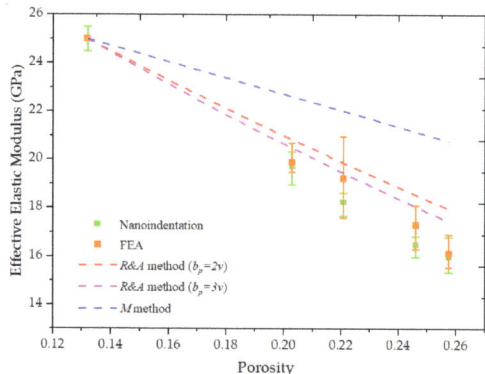

Figure 17. Effective elastic modulus degradation with porosity.

To clarify the elasticity degradation, a mechanism analysis is taken and discussed below. Micron silver atoms are held together by interaction and the elastic properties are directly related to the relative movement between atoms. Due to the mismatching of the thermal expansion coefficient of the sandwich structure, the adhesive layer is mainly subjected to shear force in the deep space environment regarding approximately monotonic loading. This may lead to lattice shift and dislocation slip in the microstructure (Figure 18).

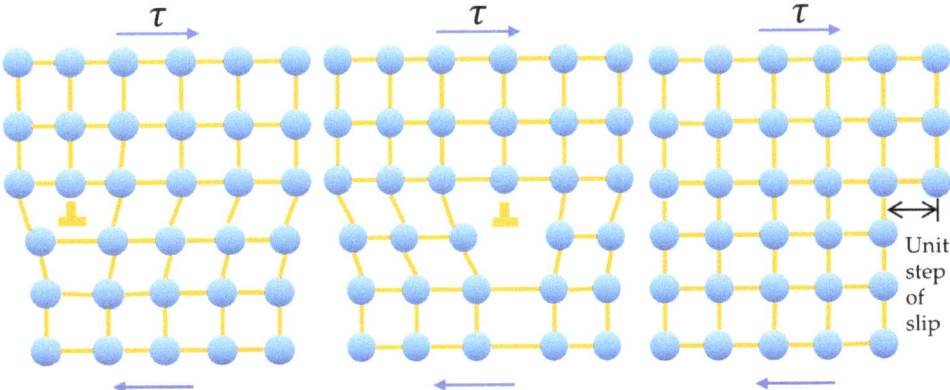

Figure 18. Dislocation from the perspective of a slip plane.

If the dislocation stops due to the concentration of micro-stress in sintered silver joints, a constrained region will be formed, where other dislocations may stop. At this time, the subsequent dislocation of the same dislocation increment occurs, piling up as shown in Figure 19. This defect can cause a decreasing number of atomic bonds; consequently, the effective elastic modulus shows degradation. Moreover, this defect can be observed as the growth of pores and dispersion of silver grains.

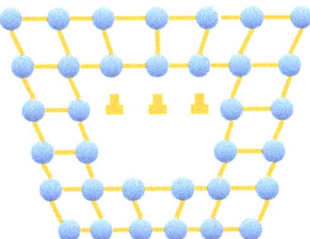

Figure 19. Ductile dislocation piling up.

5. Conclusions

This study is undertaken to provide a cost-effective modeling and simulation method for the elastic mechanical properties of pressureless micron silver sintered joints based on microstructure reconstruction and analyzing the cause of its elasticity degradation in a deep space environment. The following conclusions are drawn:

(1) A simplified statistical reconstruction method based on JQA is proposed, which takes the random morphology and distribution characteristics of the real pores into account, and applied to reconstruct the microstructure of micron silver sintered joints under thermal shock test conditions and transferred into an FEA model. The comparison of autocorrelation functions characterizing morphology features of SEM images and reconstructed models proves this method is feasible.

(2) A simulation process based on the reconstructed model is proposed to obtain the effective elastic modulus of sintered joints. Through comparing the result with that of analytical models and nanoindentation tests, the relative error (less than 15%) indicates the rationality of this method.

(3) The microstructural evolution is quantitively characterized by autocorrelation functions. In the simulative deep space environment, with increasing thermal shocks, the average micron-silver particle size and hydraulic diameter decrease, while the specific surface goes up, resulting in the increase in porosity and dispersion of silver grains.

(4) The FEA simulation results show that, due to the sandwich sample structure, the constrained sintering state makes the sintered silver joints present a slight anisotropy, which is reflected in the differences of displacement, stress and elastic modulus among X, Y, and Z directions. As thermal shock cycles increase, the material densification declines, resulting in the decreased apparent density, a larger displacement field across the local bearing surface, and more likely stress concentrations.

(5) Elasticity degradation is noticed during the thermal shock test. The mechanism analysis shows accumulated thermal shocks lead to cyclic shear force in sintered joints, which induces lattice shift and dislocation slip in the microstructure. Furthermore, dislocation piling up causes the effective elastic modulus of sintered joints to decline due to broken atomic bonds.

Author Contributions: Conceptualization, B.W.; methodology, W.G.; software, W.G.; validation, W.G. and B.W.; formal analysis, W.G.; investigation, G.F.; resources, G.F.; data curation, M.Z.; writing—original draft, W.G.; writing—review and editing, B.W.; visualization, W.G.; supervision, G.F.; funding acquisition, M.Z. All authors have read and agreed to the published version of the manuscript.

Funding: The research was funded by the Equipment Pre-research Fund Project in China, grant number 61400020105.

Acknowledgments: The authors would thank the Institute of Microelectronics, Chinese Academy of Sciences for its assistance in making samples.

Conflicts of Interest: The authors declare no conflict of interest.

References

1. Ramesham, R. Reliability of tin-lead and lead-free solders of surface mounted miniaturized passive components for extreme temperature space missions. In Proceedings of the 41st International Conference on Environmental Systems 2011 (ICES 2011), Portland, Oregon, 17–21 July 2011; GarciaBlanco, S., Ramesham, R., Eds.; Spie-Int Soc Optical Engineering: Bellingham, WA, USA, 2011; Volume 7928.
2. Li, Y.; Fu, G.; Wan, B.; Jiang, M.; Zhang, W.; Yan, X. Failure analysis of sac305 ball grid array solder joint at extremely cryogenic temperature. *Appl. Sci.* **2020**, *10*, 1951. [CrossRef]
3. Regalado, I.L.; Williams, J.J.; Joshi, S.; Dede, E.M.; Liu, Y.; Chawla, N. X-Ray Microtomography of Thermal Cycling Damage in Sintered Nano-Silver Solder Joints. *Adv. Eng. Mater.* **2019**, *21*, 1–15. [CrossRef]
4. Zhang, H.; Chen, C.; Jiu, J.; Nagao, S.; Suganuma, K. High-temperature reliability of low-temperature and pressureless micron Ag sintered joints for die attachment in high-power device. *J. Mater. Sci. Mater. Electron.* **2018**. [CrossRef]
5. Knoerr, M.; Kraft, S.; Schletz, A. Reliability assessment of sintered nano-silver die attachment for power semiconductors. In Proceedings of the 2010 12th Electronics Packaging Technology Conference (EPTC 2010), Singapore, 8–10 December 2010; pp. 56–61. [CrossRef]
6. Zhao, Z.; Zhang, H.; Zou, G.; Ren, H.; Zhuang, W.; Liu, L.; Zhou, Y.N. A Predictive Model for Thermal Conductivity of Nano-Ag Sintered Interconnect for a SiC Die. *J. Electron. Mater.* **2019**, *48*, 2811–2825. [CrossRef]
7. Zhang, H.; Wang, W.; Bai, H.; Zou, G.; Liu, L.; Peng, P.; Guo, W. Microstructural and mechanical evolution of silver sintering die attach for SiC power devices during high temperature applications. *J. Alloy. Compd.* **2019**, *774*, 487–494. [CrossRef]
8. Sun, Z.; Wang, Z.; Qian, C.; Ren, Y.; Feng, Q.; Yang, D.; Sun, B. Characterization of stochastically distributed voids in sintered nano-silver joints. In Proceedings of the 2019 20th International Conference on Thermal, Mechanical and Multi-Physics Simulation and Experiments in Microelectronics and Microsystems, (EuroSimE 2019), Hannover, Germany, 24–27 March 2019; pp. 1–6. [CrossRef]
9. Fu, S.; Mei, Y.; Li, X.; Ma, C.; Lu, G.Q. Reliability Evaluation of Multichip Phase-Leg IGBT Modules Using Pressureless Sintering of Nanosilver Paste by Power Cycling Tests. *IEEE Trans. Power Electron.* **2017**, *32*, 6049–6058. [CrossRef]
10. Okereke, M.I.; Ling, Y. A computational investigation of the effect of three-dimensional void morphology on the thermal resistance of solder thermal interface materials. *Appl. Therm. Eng.* **2018**, *142*, 346–360. [CrossRef]

11. Jiang, C.; Fan, J.; Qian, C.; Zhang, H.; Fan, X.; Guo, W.; Zhang, G. Effects of voids on mechanical and thermal properties of the die attach solder layer used in high-power LED chip-scale packages. *IEEE Trans. Compon. Packag. Manuf. Technol.* **2018**, *8*, 1254–1262. [CrossRef]
12. Zhu, L.; Xu, Y.; Zhang, J.; Liang, L.; Liu, Y. Reliability study of solder interface with voids using an irreversible cohesive zone model. In Proceedings of the 16th International Conference on Electronic Packaging Technology (ICEPT 2015), Changsha, China, 11–14 August 2015; pp. 915–920. [CrossRef]
13. Youssef, T.; Rmili, W.; Woirgard, E.; Azzopardi, S.; Vivet, N.; Martineau, D.; Meuret, R.; Le Quilliec, G.; Richard, C. Power modules die attach: A comprehensive evolution of the nanosilver sintering physical properties versus its porosity. *Microelectron. Reliab.* **2015**, *55*, 1997–2002. [CrossRef]
14. Milhet, X.; Gadaud, P.; Caccuri, V.; Bertheau, D.; Mellier, D.; Gerland, M. Influence of the Porous Microstructure on the Elastic Properties of Sintered Ag Paste as Replacement Material for Die Attachment. *J. Electron. Mater.* **2015**, *44*, 3948–3956. [CrossRef]
15. Torquato, S. Microstructural Descriptors. In *Random Heterogeneous Materials*; Springer: New York, NY, USA, 2002; pp. 23–58. [CrossRef]
16. Malmir, H.; Sahimi, M.; Jiao, Y. Higher-order correlation functions in disordered media: Computational algorithms and application to two-phase heterogeneous materials. *Phys. Rev. E* **2018**. [CrossRef]
17. Izadi, H.; Baniassadi, M.; Hormozzade, F.; Dehnavi, F.N.; Hasanabadi, A.; Memarian, H.; Soltanian-Zadeh, H. Effect of 2D Image Resolution on 3D Stochastic Reconstruction and Developing Petrophysical Trend. *Transp. Porous Media* **2018**. [CrossRef]
18. Safdari, M.; Baniassadi, M.; Garmestani, H.; Al-Haik, M.S. A modified strong-contrast expansion for estimating the effective thermal conductivity of multiphase heterogeneous materials. *J. Appl. Phys.* **2012**. [CrossRef]
19. Jiao, Y.; Stillinger, F.H.; Torquato, S. A superior descriptor of random textures and its predictive capacity. *Proc. Natl. Acad. Sci. USA* **2009**, *106*, 17634–17639. [CrossRef]
20. Izadi, H.; Baniassadi, M.; Hasanabadi, A.; Mehrgini, B.; Memarian, H.; Soltanian-Zadeh, H.; Abrinia, K. Application of full set of two point correlation functions from a pair of 2D cut sections for 3D porous media reconstruction. *J. Pet. Sci. Eng.* **2017**, *149*, 789–800. [CrossRef]
21. Tewari, A.; Gokhale, A.M.; Spowart, J.E.; Miracle, D.B. Quantitative characterization of spatial clustering in three-dimensional microstructures using two-point correlation functions. *Acta Mater.* **2004**, *52*, 307–319. [CrossRef]
22. Wang, Q.; Zhang, H.; Cai, H.; Fan, Q.; Li, G. Reconstruction of co-continuous ceramic composites three-dimensional microstructure solid model by generation-based optimization method. *Comput. Mater. Sci.* **2016**. [CrossRef]
23. Muhumthan, B. New three-dimensional modeling technique for studying porous media. *J. Colloid Interface Sci.* **1993**, *98*, 228–235. [CrossRef]
24. Bodla, K.K.; Garimella, S.V.; Murthy, J.Y. 3D reconstruction and design of porous media from thin sections. *Int. J. Heat Mass Transf.* **2014**, *73*, 250–264. [CrossRef]
25. Ramakrishnan, N.; Arunachalam, V.S. Effective elastic moduli of porous solids. *J. Mater. Sci.* **1990**, *25*, 3930–3937. [CrossRef]
26. Roberts, A.P.; Garboczi, E.J. Elastic properties of a tungsten-silver composite by reconstruction and computation. *J. Mech. Phys. Solids* **1999**, *47*, 2029–2055. [CrossRef]
27. Yeong, C.L.Y.; Torquato, S. Reconstructing Random Media I and II. *Phys. Rev. E* **1998**, *58*, 224–233. [CrossRef]
28. Dullien, F.A.L. *Porous Media: Fluid Transport and Pore Structure*; Academic Press: San Diego, CA, USA, 1979; ISBN 0122236505.
29. Nishimoto, S.; Moeini, S.A.; Ohashi, T.; Nagatomo, Y.; McCluskey, P. Novel silver die-attach technology on silver pre-sintered DBA substrates for high temperature applications. *Microelectron. Reliab.* **2018**, *87*, 232–237. [CrossRef]
30. Asoro, M.A.; Ferreira, P.J.; Kovar, D. In situ transmission electron microscopy and scanning transmission electron microscopy studies of sintering of Ag and Pt nanoparticles. *Acta Mater.* **2014**, *81*, 173–183. [CrossRef]

31. Kim, D.; Chen, C.; Pei, C.; Zhang, Z.; Nagao, S.; Suetake, A.; Sugahara, T.; Suganuma, K. Thermal shock reliability of a GaN die-attach module on DBA substrate with Ti/Ag metallization by using micron/submicron Ag sinter paste. *Jpn. J. Appl. Phys.* **2019**, *58*. [CrossRef]
32. Wang, X.; Atkinson, A. Microstructure evolution in thin zirconia films: Experimental observation and modelling. *Acta Mater.* **2011**. [CrossRef]

© 2020 by the authors. Licensee MDPI, Basel, Switzerland. This article is an open access article distributed under the terms and conditions of the Creative Commons Attribution (CC BY) license (http://creativecommons.org/licenses/by/4.0/).

Article

Accurate Real Time On-Line Estimation of State-of-Health and Remaining Useful Life of Li ion Batteries

Cher Ming Tan [1,2,3,4,*], **Preetpal Singh** [1] **and Che Chen** [1]

[1] Centre for Reliability Science and Technology, Chang Gung University, Wenhua 1st Road, Guishan Dist., Taoyuan City 33302, Taiwan; preetpalsingh96@gmail.com (P.S.); kobebrian787@gmail.com (C.C.)
[2] Center for Reliability Engineering, Ming Chi University of Technology, New Taipei City 24301, Taiwan
[3] Department of Electronic Engineering, Chang Gung University, Wenhua 1st Rd., Guishan Dist., Taoyuan City 33302, Taiwan
[4] Department of Urology, Chang Gung Memorial Hospital, Guishan, Taoyuan City 33302, Taiwan
* Correspondence: cmtan@cgu.edu.tw; Tel.: +886-3-2118800-3872

Received: 9 August 2020; Accepted: 26 September 2020; Published: 5 November 2020

Abstract: Inaccurate state-of-health (SoH) estimation of battery can lead to over-discharge as the actual depth of discharge will be deeper, or a more-than-necessary number of charges as the calculated SoC will be underestimated, depending on whether the inaccuracy in the maximum stored charge is over or under estimated. Both can lead to increased degradation of a battery. Inaccurate SoH can also lead to the continuous use of battery below 80% actual SoH that could lead to catastrophic failures. Therefore, an accurate and rapid on-line SoH estimation method for lithium ion batteries, under different operating conditions such as varying ambient temperatures and discharge rates, is important. This work develops a method for this purpose, and the method combines the electrochemistry-based electrical model and semi-empirical capacity fading model on a discharge curve of a lithium-ion battery for the estimation of its maximum stored charge capacity, and thus its state of health. The method developed produces a close form that relates SoH with the number of charge-discharge cycles as well as operating temperatures and currents, and its inverse application allows us to estimate the remaining useful life of lithium ion batteries (LiB) for a given SoH threshold level. The estimation time is less than 5 s as the combined model is a closed-form model, and hence it is suitable for real time and on-line applications.

Keywords: state of health; remaining useful life; electrochemistry based electrical model; semi-empirical capacity fading model; useful life distribution; quality and reliability assurance

1. Introduction

Electric vehicles (EV) are the focus of attention for today transportation, and their primary energy source is mainly rechargeable lithium ion batteries (LiB) due to their higher energy efficiency and longer lifetime as compared to their counterparts. However, the estimation of their health and the prediction of their remaining useful life (RUL) for EV are the major issues. In particular, the accuracy and rapid measurement of their health is a major concern.

The important health indexes for LiB have already been discussed extensively [1–5]. State-of-health (SoH), state of charge (SoC), state of energy (SoE), and state of safety (SoS) were discussed in detail for LiB. Inaccurate SoH estimation can lead to unintentional over-discharge as the actual Q_m (the remaining charge in LiB) can be a lot lower and thus 20% SoC that are ready for re-charge could actually be much lower. Such inaccuracy or uncertainty in SoH estimation can also lead to being over conservative on the user's part that increase the SoC cut off for battery charging, and lead to a higher number of charge cycles than necessary [6]. Both situations can accelerate the battery's degradation.

The unknown RUL of LiB could also result in conservative pre-mature replacement of LiB, increasing the cost of LiB in its applications [7–10]. On the other hand, if the SoH of LiB is actually already lower than 80%, prolonged usage of LiB could be problematic.

Accurate estimations of SoH and RUL are also crucial in energy storage applications. As the charging and discharging tend to be more often when LiB is used for energy storage, especially under solar and wind energy systems, large RUL with respect to charging and discharging will be important for a specific SoH threshold to justify the system cost. A sufficiently high SoH of LiB should be maintained for energy storage so that it can provide enough energy for usage when the solar energy or wind energy is no longer available. This model allows such estimation of the RUL.

Consequently, accurate estimation of SoH and prediction of RUL of LiB are crucial for LiB applications. It can be evaluated by comparing the present time performance with the ideal state performance and the battery's fresh state. There are many SoH estimation methods, and some are accurate but require complex calculation that is not suitable for real time applications. Some are not sufficiently accuracy for real applications.

The fading capacity and the increase in cell impedance of a LiB are two important factors that must be taken into consideration while estimating the SoH of the battery. The data-driven and adaptive systems are the two approaches that are generally implemented to determine the SoH of a battery. SoH estimation is performed using cycling data and parameters that affect the battery lifetime in data-driven approach. Battery's internal resistance is also used to determine SoH, but it is difficult to monitor internal resistance of the battery in the real-time scenarios. Also, deep understanding of correlation of battery operation and degradation process is required for this approach. Different designs of cells and operation conditions can have different degradation mechanisms as pointed out by Palacin [11], and hence such correlation can be difficult.

Adaptive systems approach makes use of the parameters that are sensitive to battery's degradation trend, and these parameters must be measured and examined throughout the battery operation time. However, high computational load complicates the online running of the model on a real application [12], and this is a huge drawback for adaptive systems.

Recently, Huang et al. [13] developed an online SoC and SoH estimation model for LiB. Their SoH estimation is obtained from the reciprocal of unit time voltage drop (V'), a parameter that was developed by them. However, the estimated SoH depends on the SoC at which the V' is computed, and correction factor for the given SoC is to be obtained from a correction curve, which was done via curve fitting. This can render the model applicable only to the tested LiB as the underlying physics of the correction curve is not explored. Also, their SoH computation does not depend on the charge/discharge cycle, and hence it cannot be used for the estimation of the RUL.

As SoH is a measured of the remaining maximum charge that can be stored in LiB after several cycles of charge and discharge, and with the understanding of the degradation of LiB in term of its maximum stored charge [14], it is reasonable to assume that SoH is a function of the number of charge-discharge cycle, the discharge current, and temperature. Palacin [15] provided a good overview of the main ageing mechanism in LiB, and she showed that the combination of high current and high temperature are most detrimental to the degradation of LiB.

Liu et al. [16] proposed a semi-empirical capacity fading (SECF) model based on the above-mentioned factors, and Preetpal et al. [17] recently demonstrated the successful use of this model for the estimation of SoH of a set of LiB at different charge-discharge cycles, and the estimation error was less than 2.5% when the extrapolation went beyond 300 cycles of operation. However, their work did not consider the effect of temperature and discharge rate in the estimation of SoH. In this work, we extend our investigation of the model in estimating SoH of LiB under different temperatures and discharge rates. As the equation is in a compact closed form, it can be implemented easily in real time, and we could also able to use the equation to determine the cycle at a specific SoH threshold, such as 80%, making prediction of its RUL possible in both EV and energy storage applications. A noteworthy point is that the depth of discharge of LiB during operation is not considered in this work although it is known to affect the SoH of LiB. This is because of its complex degradation mechanism for LiB, and it will be treated in the future work.

To evaluate the accuracy of the SoH estimation using the SECF model, we use the electrochemistry-based electrical model (ECBE) model [18] as reference. The ECBE model is based on the first principle of electrochemistry, which is unlike the previously reported models, which employ the best-fit techniques. ECBE model results are also verified using the electrochemical impedance spectrometer (EIS) results. In fact, EIS is also used to understand the various aging mechanisms using electrical models, but the method can be done only offline in frequency domain. On the other hand, ECBE allows the performance of each component inside LiB be determined real time through its discharging curve non-destructively (i.e., terminal voltage vs. time during discharge), making it suitable for field applications. Although this ECBE model is accurate, the computation time to obtain the values of the model parameters is too excessive, which renders its unsuitability for in-situ real time SoH estimation.

The paper is organized as follows. The experimental settings and approach are given in Section 2, and a brief introduction of the semi-empirical capacity fading model is shown in the subsequent section. The applications of the SECF model for cases of different discharge rate and temperature as well as its SoH estimation accuracy are presented next. Conclusions and future works are given in the last section.

2. Materials and Methods

4 Panasonic SANYO UR18650E lithium-ion batteries [19] are tested using Bio-Logic BCS-815, which is an 8-channel tester and support 15 A per channel. It is used for charging and discharging of the batteries with test data collection under room temperature, which is approximately 25 °C. The sampling frequency of the data collection is 1 Hz. The specification of the batteries used here are shown in Table 1 as obtained from the manufacturer.

Table 1. Samsung 18,650 battery specification provided by manufacturer [8].

Battery Characteristics	
Type	Cylindrical
Chemical system	NMC
Nominal voltage	3.62 V
Typical capacity	2150 mAh
Cut-off voltage Charging	4.2 V
Discharging	2.75 V
Dimensions(mm)	18.4 × 65
Approx. weight	44.5 g

The batteries are tested under different test conditions as shown in Table 2. Only one battery is tested under each condition due to time and resources limitations. However, as the model employed here has been verified in the work by Preetpal et al. [17] on different charge-discharge cycles for LiB, the results obtained in this work are presented with confidence despite the small sample size. Verification of the model using another 3 cells from the same batch will be performed later. The purpose of this work is to examine the capability and accuracy of this SECF model when ambient temperature and discharge current are considered. More batteries will be tested when resource becomes available.

Table 2. Test conditions for the batteries.

Cell Name	Test	Temperature (°C)	Discharge Current
A	1	55	1 C
B	2	55	3 C
C	3	25	1 C
D	4	25	3 C

In the experiments, we use CC-CV method for charging the batteries. 1 A constant current (as recommend from the battery's manufacturer datasheet) is used to charge the batteries to its cut-off voltage, and the charging is done with constant voltage until the current drops to 100 mA. When the battery is fully charged, a rest time of 30 min is kept before discharging. One or 3 C-rate is implemented to discharge the batteries until the cut-off voltage of 4.2 V is reached. Figure 1 shows the typical time progression of the terminal voltage of a LiB during charging and discharging. One can see that the voltage is higher when LiB is charging at higher temperature. This is because cell impedance will be higher at higher temperature. The terminal voltage can be modelled as the constant current multiples the cell impedance as the charging is done using CC-CV, and hence a higher terminal voltage results. In fact, the cell impedance of a LiB has been employed to monitor the temperature of LiB by Beelen et al. [20].

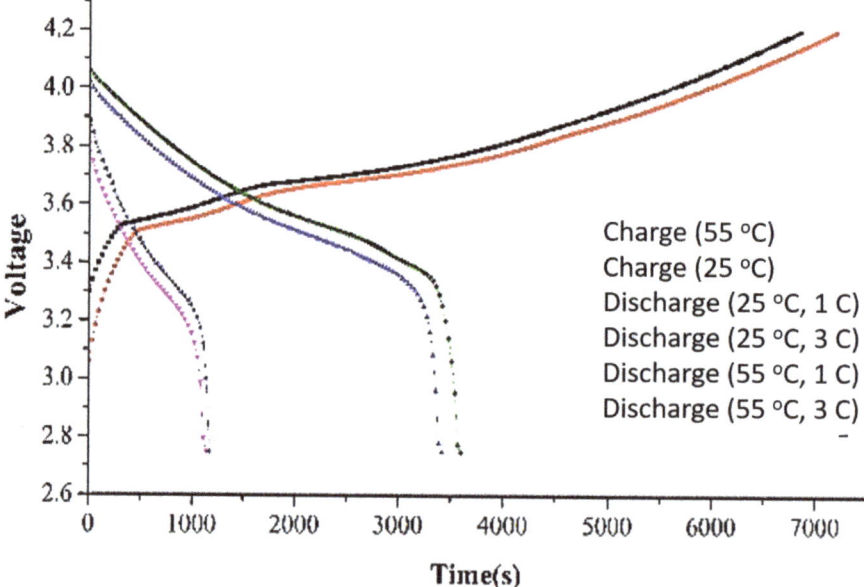

Figure 1. Variation of terminal voltage recorded during discharge period for 1 and 3 C-rate at different temperature.

The discharge curves of all the LiB are analyzed using the ECBE model to calculate the maximum capacity of the cells (Q_m). This Q_m is used to compute the corresponding SOH using Equation (1) where $Q_{max(fresh)}$ is the Q_m after the first discharge cycle, and $Q_{max(aged)}$ is the Q_m after the subsequent cycles. The SoH computed is termed as Experimental SoH when the Q_ms are determined using ECBE model.

$$\text{SoH} = \frac{Q_{max(aged)}}{Q_{max(fresh)}} \qquad (1)$$

The SECF model used in this work is shown in Equation (2) [16]. The SoH determined using Equation (2) is termed as Estimated SoH. The parameter k_1 accounts for the capacity losses that increase rapidly during the conditions of cycling at high temperature, and k_2 is a factor to account for capacity losses under the normal conditions of cycling. k_3 accounts for the capacity loss due to C-rate [21].

$$\text{SoH} = 1 - \left(0.5 * k_1 N^2 + k_2 N\right) - \frac{k_3}{Q_{max(fresh)}} i \qquad (2)$$

Here, N represents the number of charge-discharge cycles the battery experienced at the time of this SoH calculation and i is discharging current.

These parameters (k_1, k_2, and k_3) can be extracted by substituting the Experimental SoH obtained from Equation (1) into Equation (2) at 3 different cycles (the exact cycles can be seen in Table 3 later in order to ensure the largest cycle chosen among the 3 cycles is half of the cycle at which the SoH of the LiB is around 80%) as shown in Figure 2. The extracted parameters are then used to calculate the Estimated SoH in this work. Our experimental procedure is depicted in Figure 3. There are large fluctuations or non-linearity observed in Qm over the first few cycles, and this non-linearity is higher at lower temperatures, which is also observed by other researchers [14], thus the first 50 cycles are not used for the k values extraction.

Table 3. k values of the semi-empirical capacity fading (SECF) model obtained at four test conditions.

Temperature (°C)	C-Rate	3 Cycles for the Extraction of k Values	k_1	k_2	k_3
25	1	100, 200, 300	0	0.000283	0.0027
25	3	100, 200, 300	0	0.0000599	0.0101
55	1	100, 200, 300	0	0.000354	0
55	3	75, 125, 250	0	0.00045	1.43×10^{-3}

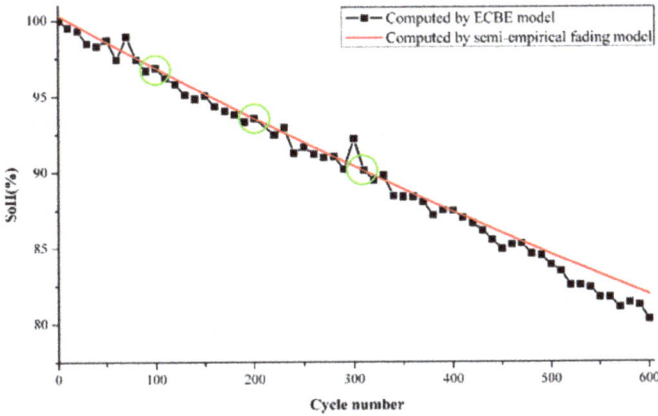

Figure 2. State-of-health (SoH) estimation results from ECBE model (black curve) and semi-empirical fading model (red curve) for battery tested at 55 °C and 1 C discharge current. Green circle represents the points used to find the k values.

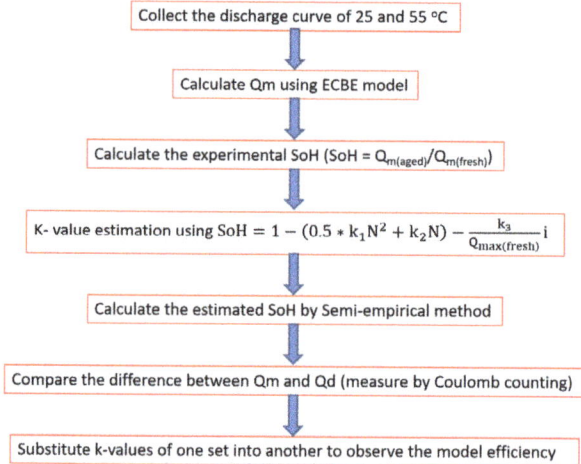

Figure 3. Flowchart for error estimation between experimental and estimated SoH.

3. Results

To demonstrate the effect of varying C-rate at constant temperature and the effect of varying temperatures at a constant C-rate on SoH estimation accuracy of the SECF model, we divide our results into four sections as shown in Figure 4. In Figure 4, A, B, C and D represent cases where k values are obtained from batteries tested at 1 C_55 °C, 3 C_55 °C, 1 C_25 °C, and 3 C_55 °C, respectively. On the other hand, A' and B' represent the SoH estimation errors for cells A and B using k values obtained from cell A. C* and D* represent the SoH estimation errors for cells C and D using k values obtained from cell A. Similarly, A" and B" are the SoH estimation errors for cells A and B using k values obtained from cell B. Others follow the same notations.

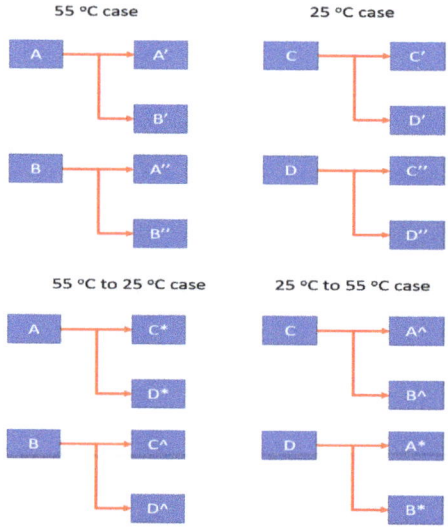

Figure 4. Notations used in the analysis of test results.

3.1. Effect of C-Rate at Constant Temperature Conditions

3.1.1. SOH Estimation for Batteries under Different Discharge C-Rate at 55 °C

Experimental SoH is calculated for each cell tested for different number of cycles at two C-rates and 55 °C, and the results are shown in Table 4. SoH decreases with the cycle number and the rate of decrease of SoH is higher for battery discharged at 3 C rate as expected. The sample with 1 C-rate reaches the cut-off point of around 80% SoH after 600 cycles, while the sample with 3 C-rate reaches the cut-off point before 500 cycles.

Table 4. Experimental SoH at 55 °C based on the Qm computed from the ECBE model.

Cycle Number	1 C	3 C
0	100	100
100	96.86	95.24
200	93.55	91.29
300	90.21	87.68
400	87.46	82.98
500	83.88	79.67
600	80.28	-

The calculated SoH values shown in Table 4 are used to obtain the k's values of the SECF model, and the values are shown in Table 5. The very small negative values of k_1 is likely due to fitting approximation error and hence they are set to zero. In fact, k_1 value should be zero as our ambient temperature is not high as expected from the work by [21].

Table 5. The k-values of 1 and 3 C-rate, respectively, at 55 °C.

	k_1	k_2	k_3
1 C-rate of 55 °C (A)	-1.61478×10^{-7}	3.5497×10^{-4}	0
3 C-rate of 55 °C (B)	-3.7322×10^{-7}	4.5086×10^{-4}	1.43376×10^{-3}

To verify the accuracy of the SECF model, the k-values obtained from one battery is applied to obtain the SoH of same LiB. The percentage error is then evaluated by comparing the estimated SoH with the experimental SoH. Figure 2 shows such comparison and the result is satisfactory.

To study the effect of C-rate on the estimation accuracy, the k-values obtained from one battery is applied to obtain the SoH of other battery tested under same temperature but at other C-rate. Figure 5 summarizes the percentage errors at other discharge conditions. It is to be noted that the maximum percentage error on estimating SoH using semi-empirical model is around 3.14% using the k-values of battery tested at 1 C and applied to battery tested at 3 C, and the percentage error is around 2.11% for the reverse case. Both errors are within the acceptable limit of 10%.

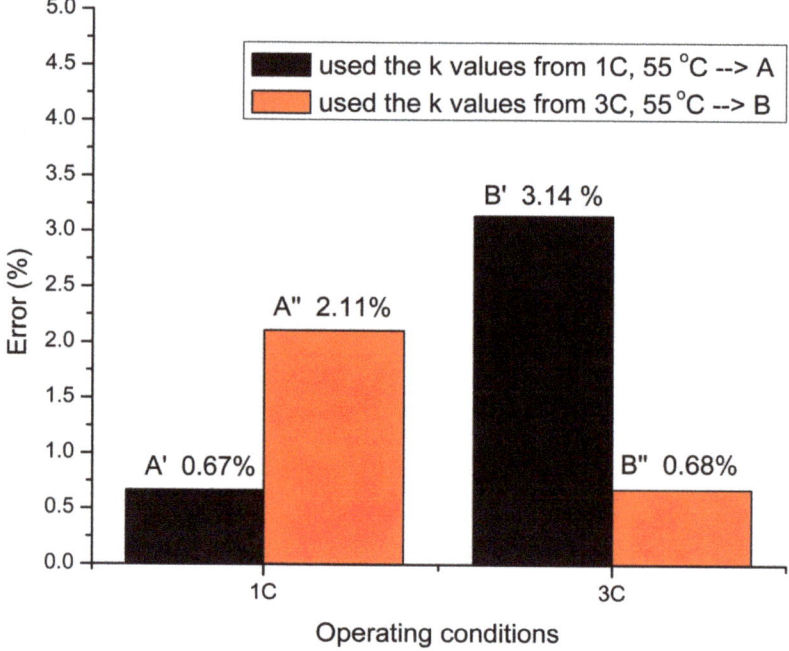

Figure 5. The estimation error between different discharge current under 55 °C. % value on top of the bar represents the % error in estimation.

From Figure 5, we can see that the estimation error of using the SECF model is sufficiently accurate for predicting the SoH of its own cell at larger cycle number, and this has been elaborated in the work by Preetpal et al. [17]. We can also see that it is fairly accurate when it is applied to the cells at different C rates. In comparing the percentage errors of A″ and B′, although the different in the C rate is the same, the largest percentage error is observed for the case of B′. This is expected because of the largest temperature change of the cells from the reference cell to the cell of interest in the case of B′. We will discuss this in more detail later. Hence, we see that both the different in the discharge rate and the cell temperature can affect the estimation accuracy.

In terms of the selection of the cycles to estimate the k's values, we select a different set of the three cycles for the SoH estimation of battery tested at 55 °C and 1 C rate. Table 6 shows the comparison, and one can see that the effect of the selection on the percentage error in SoH estimation is not significant.

Table 6. Percentage estimation error of SoH for a different set of three cycles in the k's values extraction.

Cycles Selected for k's Values Extraction	% Estimation Error
100, 200, 300	3.14
125, 225, 325	3.68
150, 250, 350	3.52

3.1.2. SoH Estimation for Batteries at Different C-Rates under 25 °C Ambient

We perform a similar study as in the previous section to estimate SoH for batteries tested at two C-rates and 25 °C, and the results are shown in Tables 7 and 8. Batteries tested under 25 °C shows lower degradation rate at both 1 C-rate and 3 C-rate when compared with batteries tested at 55 °C as expected. Also, the degradation rate is slower for 1 C-rate as expected. Sample using 1 C-rate still have 89.76% SoH after 800 cycles, while the 3 C-rate sample reaches the cut-off point at around 700 cycles.

Table 7. Calculated SoH at 25 °C based on the Qm computed from the ECBE model.

Cycle Number	1 C	3 C
1	100%	100%
100	97.38%	96.28%
200	94.91%	93.00%
300	94.26%	91.30%
400	93.14%	88.21%
500	92.48%	84.48%
600	91.88%	83.80%
700	90.81%	80.41%
800	89.76%	-

Table 8. The k-values of 1 and 3 C-rate respectively at 25 °C. Negative value of k_1 is set to zero.

k-Values	k_1	k_2	k_3
1 C-rate of 25 °C (C)	-8.9785×10^{-8}	2.8312×10^{-4}	0.0027
3 C-rate of 25 °C (D)	8.3792×10^{-7}	5.9967×10^{-5}	0.0101

Again, the k's values obtained from one battery is applied to obtain the SoH of same battery as well as other battery tested at other C-rate and the error percentage is computed between the Experimental and Estimated SoH. The percentage error results for estimating SoH are shown in Figure 6.

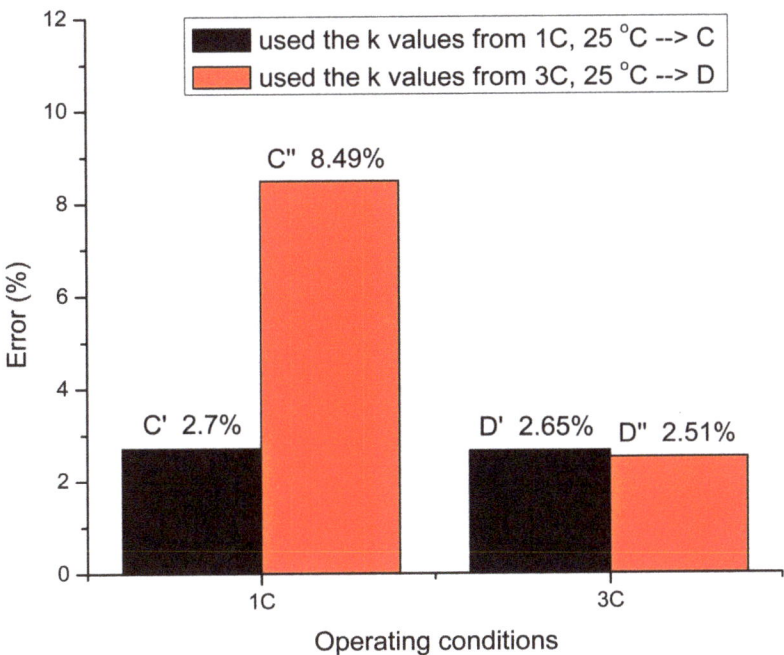

Figure 6. The estimation error between 1 and 3 C-rate at 25 °C. The % value on top of each bar represent the % error of estimation.

In comparing the % errors for C′ and A′, and that for D′ and B′, we can see that the % error is larger for the case of 25 °C. Such larger error is due to the inaccuracy of the temperature of 25 °C as the testing was done in the laboratory ambient, and the ambient temperature has a small fluctuation over the span of measurement period. The temperature varies between 24 to 26 °C during our test.

In Figure 6, the largest percentage error obtained is around 8.49 % for the case when k's values from battery tested at 3 C is applied to obtain SoH for battery tested at 1 C under 25 °C. In comparison to D′, although the difference in the C rate between the cells are the same, the large % error for the case of C″ is expected to be attributed by the temperature difference between the cell of interest and the reference cell, which is approximately 8 °C as can be seen in Table 9. Table 9 shows the temperatures of the LiB when discharged at different C rates at different temperatures.

Table 9. Cell temperature when discharged at different C rate under two ambient conditions.

	1 C Rate	3 C Rate
25 °C Ambient	34.86 °C	42.32 °C
55 °C Ambient	58 °C	63 °C

The increase in the cell temperature as shown in Table 9 is expected according to the work by Huang et al. [22], and the rise in temperature during discharge depends on the Peltier heat as a result of the entropy change, the discharge current, heat capacity and weight of a battery as well as its thermal design and the ambient temperature [22]. As the different C rate and cell temperature can affect the estimation accuracy, we further investigate the effect of temperature on the accuracy of the estimation using the SECF model.

3.2. Effect of Temperature

3.2.1. SOH Estimation for Batteries at 25 °C using 55 °C Battery's Parameter Values (55 °C to 25 °C Case)

Similar to our previous approach, we use the k values from the batteries with similar C-rate but apply them to SoH estimation of cells discharged at different temperatures. First, SoH estimation is performed for batteries tested at 25 °C and the k values obtained from battery tested at 55 °C are used for the SoH estimation, and the results are shown in Figure 6.

It can be observed from Figure 7 that the highest SoH estimation percentage error of 8.4% is obtained for the case when k values of battery tested at 3 C rate and 55 °C is used to estimate the SoH for battery tested at 1 C rate and 25 °C. This is due to the largest cell temperature difference between the cells, which is approximately 28.14 °C in this case as can be seen in Table 9. The smallest estimation error for the estimation of battery tested at 3 C and 25 °C is corresponding to the small difference in cell temperature for cell discharge at 1 C and 55 °C and cell discharge at 3 C 25 °C. This finding supports our postulation that cell temperature difference is an important factor in the estimation accuracy.

Appl. Sci. **2020**, *10*, 7836

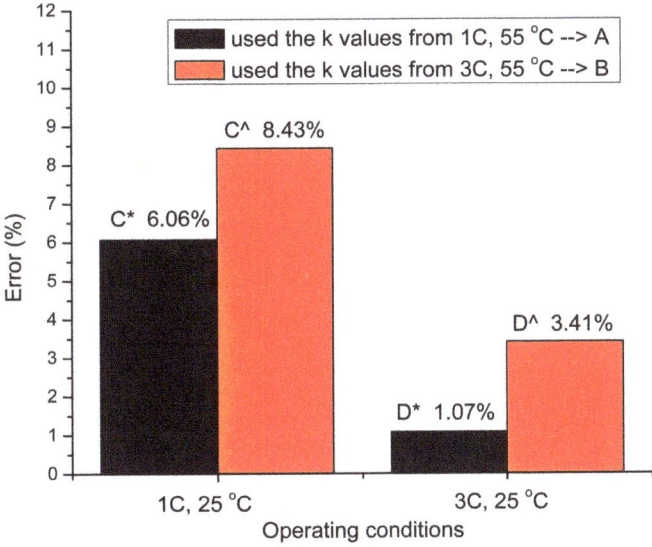

Figure 7. Estimate % error using k's values from 55 °C.

3.2.2. SOH Estimation for Batteries at 55 °C using 25 °C Battery's Parameter Values (25 °C to 55 °C Case)

Now we study the opposite situation where estimation is made using parameters values from low temperature to high temperature, and the results are shown in Figure 8.

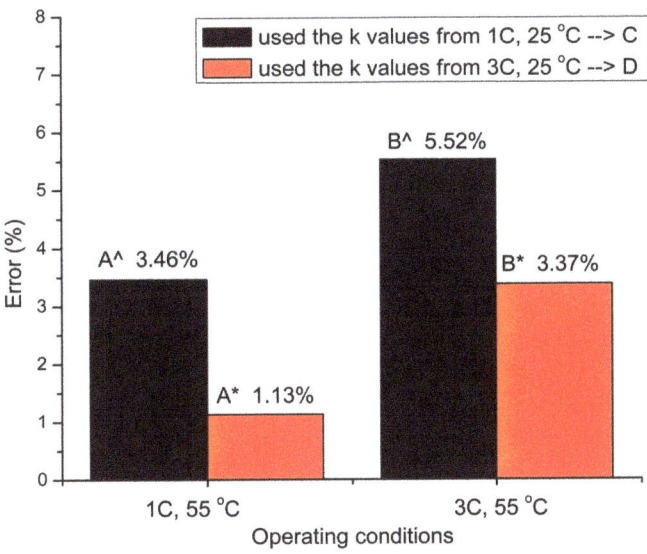

Figure 8. Estimation % error using k values from 25 °C.

From Figure 8, we can see that the smallest percentage error is the case where the parameters values from 3 C and 25 °C is applied to 1 C at 55 °C. This is expected with our postulation where the cell temperature difference between the two cells is only 15.68 °C in this case. Likewise, the largest percentage error is observed for the case where the parameters values from 1 C at 25 °C is applied to 3 C at 55 °C and the temperature difference is as high as 28.14 °C.

From the above investigation, the interaction of discharge current and cell temperature needs to be identified. The statistical design of experiment (DoE) method, a commonly used method in industry for studying the effect of multiple factors, is thus employed. The detail of the DoE method can be found in reference [23]. We summarized the k values at the four test conditions as shown in Table 3. The interaction of the discharge current and temperature as reflected in the k values can be plotted as shown in Figures 9 and 10, as part of the DoE analysis. Using these values and applying the 2^2 factorial design analysis, we obtain the equations for k_2 and k_3 as given in Equations (3) and (4). k_1 is zero as our experiments do not involve high temperatures.

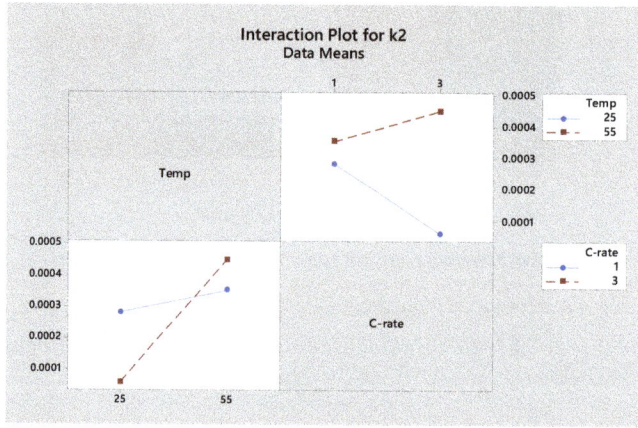

Figure 9. Interaction of current density and temperature in the k_2 value.

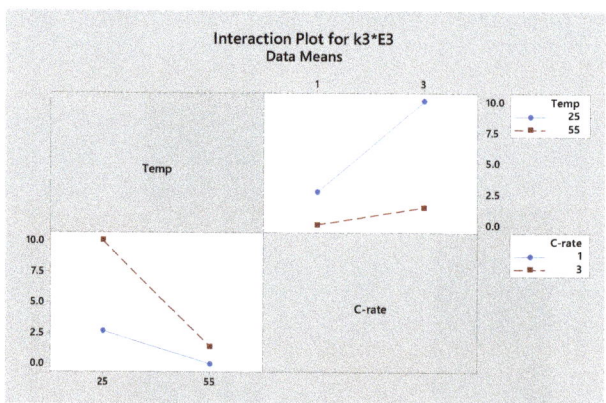

Figure 10. Interaction of current density and temperature in the k_3 value.

From the DoE analysis, we obtain the following equations for k_2 and k_3.

$$k_2 = 0.000287 - 0.000115\,A - 0.000080\,B - 0.000032\,A \times B \tag{3}$$

$$k_3 = 0.003557 + 0.002207\,A + 0.002843\,B + 0.001493\,A \times B \tag{4}$$

where

$$A = (\text{Temperature (°C)} - 40\ °C)/15\ °C \qquad (5)$$

and

$$B = C\ \text{rate} - 2 \qquad (6)$$

Using Equations (3) and (4), we can estimate the SoH via Equation (2), and the percentage errors in the SoH estimation with respect to the Experimental SoH can be seen in Table 10. One can see that with the inclusion of the interaction of discharge current and temperature, the % errors in the SoH estimation are consistently less than 1%.

Table 10. Comparison of the % errors in SoH estimation.

Test Condition	SoH (Using ECBE)	% Error in SoH Estimation Using Its Own k Values	% Error in SoH Estimation with k Values Computed from Equations (3) and (4)
25 °C_1 C	89.76(after 800 cycles)	2.7	0.89
25 °C_3 C	80.41(after 300 cycles)	2.51	0.77
55 °C_1 C	80.29(after 600 cycles)	0.67	0.59
55 °C_3 C	79.68(after 500 cycles)	0.68	0.41

We also extend our equations to the cases of 0.5 C and 5 C at 55 °C. The % errors in the estimation are shown in Table 11, and one see that the % error is small even for the case of 5 C where cell heating is serious. As 5C at 55 °C is likely to be the highest cell temperature for LiB, this implies that k_1 is always zero for the case of LiB

Table 11. % error in SoH estimation for the cases of 0.5 and 5 C at 55 °C

Test Condition	SoH (Using ECBE)	% Error in SoH Estimation	% Error in SoH Estimation with Generalized k Values
55 °C_0.5 C	80.27(After 600 cycles)	0.97	0.84
55 °C_5 C	76.08(After 200 cycles)	5.47	0.61

To verify the proposed model, and for the purpose of reproducibility, three other batteries from the same batch are tested at 55 °C_5 C and Table 12 shows the % errors in the SoH estimation estimated using Equations (3) and (4). The test condition is chosen to save time as the battery degradation is fastest in such harsh condition. Batteries are tested for 100 cycles. The results in Table 12 shows the good accuracy of the proposed method.

Table 12. Verification of the model for additional three more batteries tested at 55 °C and 5 C discharging current.

Sample #	SoH (Using ECBE)	SoH Estimation Using Our Generalized k Values	% Error in SoH Estimation with Generalized k Values
1	89.06(After 100 cycles)	88.24	0.92
2	91.39(After 100 cycles)	90.72	0.73
3	90.66(After 100 cycles)	89.94	0.79

4. Applications of the SECF Model

From the previous analysis, we can see that with the equation of SoH given by Equation (2) and the k values computed using Equations (3) and (4), one can estimate the SoH of LiB after different charge-discharge cycle. Conversely, one can also determine the cycles where the SoH will reach a certain value due to its closed form. For example, if the LiB cell is operating under 2 C discharge rate at 40 °C, one can have the following equation, with i = 2C, k_2 = 0.00028, k_3 = 0.003557, and $Q_{max(fresh)}$ = 1.9463 Ah. Upon solving it, the n value is 689, which will be the cycles where the SoH reaches 80%, and thus the useful life of the LiB cell at different operating condition can be determined.

$$80 = 1 - (k_2 N) - \frac{k_3}{Q_{max(fresh)}} i \qquad (7)$$

The above example is an application of the SECF model for the lifetime prediction of a fresh LiB. This model can also be used for the estimation of the RUL as follows.

Given the k_2 and k_3 values of a LiB as in this work, and assuming the LiB is used in 1 C discharge rate and has undergone 100 cycles at 30 °C, the LiB is now to be used in 1.5 C discharge rate at the same temperature of 30 °C. If this LiB has been through 20 cycles under 1.5 C discharge rate, we can use our SECF model to compute the RUL of the LiB under this 1.5 C discharge rate as follows.

For the SoH of the LiB after the 100 cycles @1C discharge rate, A = −10/15 and B = −1 according to Equations (5) and (6), and with the values of k_2 and k_3 as computed using Equations (3) and (4), we obtain the SoH after 100 cycles to be 96.25%, using Equation (2). This is the SoH of the LiB at the start of the next application of 1.5 C. The equivalent cycle for the LiB to reach this SoH at 1.5 C and 30 °C can be computed from the model by having the B = −0.5. A will not change as the temperature remains the same. With this new B, we have new values of k_2 and k_3 and the equivalent cycle is found to be 88 cycles ($N_{equivalent}$).

We can also compute the total number of cycles at 1.5 C and 30 °C to reach 80% SoH, denoted as N_{total}. From our SECF model, the value is found to be 769 cycles. Since the LiB has been through 20 cycles under 1.5 C, the RUL of the LiB will be $N_{total} - N_{equivalent} - 20 = 769 - 88 - 20 = 661$ cycles.

Going forward, if the discharge current varies, one can use the average current, expressed in terms of C rate to compute the B value. If the temperature is varying, one can use the average temperature to compute the A value. Then, using Equations (2)–(4), the SoH or the RUL can be determined. However, this is to be verified as our future work.

5. Future Works

While the SECF model is promising as shown in this work, the sample size used in this work is too small. In order to increase the confidence of this model, larger sample size and with LiB from different manufacturers as well as different type of LiB are to be used. This requires large resources and long test time. Collaboration with other research organizations on LiB is desired.

As mentioned in the introduction, the proposed model does not include the effect of the depth of discharge, which is known to affect SoH. This is one of the limitations of the proposed model and the inclusion of the depth of discharge in the model will also be our future work. Another limitation is the possible change of the values of the k parameters when the SoH becomes low as the internal structure of the LiB could have changed so significantly that a new set of k parameters are to be extracted from the ECBE model which has shown to be accurate even at low SoH [14]. Such investigation will also be our future work.

Besides the above-mentioned applications, other applications of SECF model are possible and yet to be verified. One of them is the development of acceleration model of LiB lifetime. If we can perform the same four set of test conditions at higher temperature and discharge rate, the k values can be obtained in a short time, and hence, the complete SoH estimation equations can be obtained and applied to compute the lifetime of LiB cell in other operating conditions, in the same way as in the above examples, for a given heat capacity, weight, and thermal design of the cells. Thus, battery manufacturers can simply provide a lifetime model for each batch of their batteries for their users. Also, as k values determined the lifetime of LiB, statistical distribution of the k values from a set of LiB samples in a production batch can also provide information on the lifetime distribution of the batch of LiB, and hence a process control methodology can be developed.

6. Conclusions

The requirement of accurate online estimation of LiB battery capacity and its remaining useful life (RUL) are important. In this work, we applied ECBE model and semi-empirical capacity fading (SECF) model to estimate SoH and its remaining useful life after a LiB is operated for different charge-discharge cycles. ECBE model is only used for initial SoH calculation twice to determine the parameters in the SECF model, which can then be used for the SoH and RUL estimation of cells for their subsequent charge-discharge cycles.

We demonstrate that SECF model for SoH estimation can provide accurate SoH estimation, using the ECBE model as reference, regardless of the discharge current and ambient temperature. The discharge rate considered is from 0.5 C to 5 C, and the ambient temperature is from 25 °C to 55 °C. As the model is in compact closed form, it can be used for the lifetime prediction of LiB, and this could provide a good method for LiB acceleration test so that the SECF model parameters can be determined for its lifetime estimation of LiB in other operating conditions in a short time. Also, with this closed form, this model is practical for on-line real time SoH estimation, as the estimation time after each discharge cycle is less than 5 s using a personal computer with capability of 8 GB RAM and Intel Core i5 processor.

Author Contributions: C.M.T. conceived and designed the research; P.S., C.C., and C.M.T. analyzed the testing data; P.S., C.C., and C.M.T. wrote the paper. All authors have read and agreed to the published version of the manuscript.

Funding: This research was funded by Chang Gung University and Ming Chi University of Technology, Taiwan and The APC was funded by Chang Gung University, Taiwan.

Acknowledgments: The authors would like to acknowledge the support of Centre of Reliability Science and Technologies Lab, Chang Gung University and Center on Reliability Engineering, Ming Chi University of Technology for providing valuable equipment and support for smooth conduct of experiments.

Conflicts of Interest: The authors declare no conflict of interest.

References

1. Cabrera-Castillo, E.; Niedermeier, F.; Jossen, A. Calculation of the State of Safety (SOS) for Lithium Ion Batteries. *J. Power Sources* **2016**, *324*, 509–520. [CrossRef]
2. Casimir, A.; Zhang, H.; Ogoke, O.; Amine, J.C.; Lu, J.; Wu, G. Silicon-Based Anodes for Lithium-Ion Batteries: Effectiveness of Materials Synthesis and Electrode Preparation. *Nano Energy* **2016**, *27*, 359–376. [CrossRef]
3. Liu, X.; Wu, J.; Zhang, C.; Chen, Z. A Method for State of Energy Estimation of Lithium-Ion Batteries at Dynamic Currents and Temperatures. *J. Power Sources* **2014**, *270*, 151–157. [CrossRef]
4. Mamadou, K.; Delaille, A.; Lemaire-Potteau, E.; Bultel, Y. The State-of-Energy: A New Criterion for the Energetic Performances Evaluation of Electrochemical Storage Devices. *ECS Trans.* **2010**, *25*, 105–112. [CrossRef]
5. Moo, C.S.; Ng, K.S.; Chen, Y.P.; Hsieh, Y.C. State-of-Charge Estimation with Open-Circuit-Voltage for Lead-Acid Batteries. In Proceedings of the Fourth Power Conversion Conference-NAGOYA, Nagoya, Japan, 2–5 April 2007; pp. 758–762. [CrossRef]
6. Baumhöfer, T.; Brühl, M.; Rothgang, S.; Sauer, D.U. Production Caused Variation in Capacity Aging Trend and Correlation to Initial Cell Performance. *J. Power Sources* **2014**, *247*, 332–338. [CrossRef]
7. Wang, R.; Feng, H. Lithium-Ion Batteries Remaining Useful Life Prediction Using Wiener Process and Unscented Particle Filter. *J. Power Electron.* **2020**, *20*, 270–278. [CrossRef]
8. Zhang, L.; Mu, Z.; Sun, C. Remaining Useful Life Prediction for Lithium-Ion Batteries Based on Exponential Model and Particle Filter. *IEEE Access* **2018**, *6*, 17729–17740. [CrossRef]
9. Ungurean, L.; Cârstoiu, G.; Micea, M.V.; Groza, V. Battery State of Health Estimation: A Structured Review of Models, Methods and Commercial Devices. *Int. J. Energy Res.* **2017**, *41*, 151–181. [CrossRef]
10. Dong, G.; Chen, Z.; Wei, J.; Ling, Q. Battery Health Prognosis Using Brownian Motion Modeling and Particle Filtering. *IEEE Trans. Ind. Electron.* **2018**, *65*, 8646–8655. [CrossRef]

11. Leng, F.; Wei, Z.; Tan, C.M.; Yazami, R. Hierarchical Degradation Processes in Lithium-Ion Batteries during Ageing. *Electrochim. Acta* **2017**, *256*, 52–62. [CrossRef]
12. Lu, L.; Han, X.; Li, J.; Hua, J.; Ouyang, M. A Review on the Key Issues for Lithium-Ion Battery Management in Electric Vehicles. *J. Power Sources* **2013**, *226*, 272–288. [CrossRef]
13. Huang, S.-C.; Tseng, K.-H.; Liang, J.-W.; Chang, C.-L.; Pecht, M. An Online SOC and SOH Estimation Model for Lithium-Ion Batteries. *Energies* **2017**, *10*, 512. [CrossRef]
14. Leng, F.; Tan, C.M.; Yazami, R.; Le, M.D. A Practical Framework of Electrical Based Online State-of-Charge Estimation of Lithium Ion Batteries. *J. Power Sources* **2014**, *255*, 423–430. [CrossRef]
15. Palacín, M.R. Understanding Ageing in Li-Ion Batteries: A Chemical Issue. *Chem. Soc. Rev.* **2018**, *47*, 4924–4933. [CrossRef] [PubMed]
16. Liu, Z.; Tan, C.; Leng, F. A Reliability-Based Design Concept for Lithium-Ion Battery Pack in Electric Vehicles. *Reliab. Eng. Syst. Saf.* **2015**, *134*, 169–177. [CrossRef]
17. Singh, P.; Chen, C.; Tan, C.M.; Huang, S.-C. Semi-Empirical Capacity Fading Model for SoH Estimation of Li-Ion Batteries. *Appl. Sci.* **2019**, *9*, 3012. [CrossRef]
18. Leng, F.; Tan, C.M.; Yazami, R.; Maher, K.; Wang, R. Quality decision for overcharged Li-Ion battery from reliability and safety perspective. In *Theory and Practice of Quality and Reliability Engineering in Asia Industry*; Springer: Singapore, 2017; pp. 223–232.
19. Specification of Product Specification of Product Specification of Product Specification of Product for Lithium-Ion Rechargeable Cell; 2010. Available online: http://www.batteryspace.com/prod-specs/4869.pdf (accessed on 28 September 2020).
20. Beelen, H.; Mundaragi Shivakumar, K.; Raijmakers, L.; Donkers, M.C.F.; Bergveld, H.J. Towards Impedance-based Temperature Estimation for Li-ion Battery Packs. *Int. J. Energy Res.* **2020**, *44*, 2889–2908. [CrossRef]
21. Ramadass, P.; Haran, B.; White, R.; Popov, B.N. Mathematical Modeling of the Capacity Fade of Li-Ion Cells. *J. Power Sources* **2003**, *123*, 230–240. [CrossRef]
22. Huang, Q.; Yan, M.; Jiang, Z. Thermal Study on Single Electrodes in Lithium-Ion Battery. *J. Power Sources* **2006**, *156*, 541–546. [CrossRef]
23. Amazon.com: Design and Analysis of Experiments (9781118146927): Montgomery, Douglas C.: Books. Available online: https://www.amazon.com/Design-Analysis-Experiments-Douglas-Montgomery/dp/1118146921 (accessed on 16 April 2020).

Publisher's Note: MDPI stays neutral with regard to jurisdictional claims in published maps and institutional affiliations.

© 2020 by the authors. Licensee MDPI, Basel, Switzerland. This article is an open access article distributed under the terms and conditions of the Creative Commons Attribution (CC BY) license (http://creativecommons.org/licenses/by/4.0/).

Article

Reliability Analysis Based on a Gamma-Gaussian Deconvolution Degradation Modeling with Measurement Error

Luis Alberto Rodríguez-Picón [1,*], Luis Carlos Méndez-González [1], Roberto Romero-López [1], Iván J. C. Pérez-Olguín [1], Manuel Iván Rodríguez-Borbón [1] and Delia Julieta Valles-Rosales [2]

[1] Department of Industrial Engineering and Manufacturing, Institute of Engineering and Technology, Autonomous University of Ciudad Juárez, Ciudad Juárez 32310, Mexico; luis.mendez@uacj.mx (L.C.M.-G.); rromero@uacj.mx (R.R.-L.); ivan.perez@uacj.mx (I.J.C.P.-O.); ivan.rodriguez@uacj.mx (M.I.R.-B.)

[2] Department of Industrial Engineering, New Mexico State University, Las Cruces, NM 30001, USA; dvalles@nmsu.edu

* Correspondence: luis.picon@uacj.mx

Abstract: In most degradation tests, the measuring processes is affected by several conditions that may cause variation in the observed measures. As the measuring process is inherent to the degradation testing, it is important to establish schemes that define a certain level of permissible measurement error such that a robust reliability estimation can be obtained. In this article, an approach to deal with measurement error in degradation processes is proposed, the method focuses on studying the effect of such error in the reliability assessment. This approach considers that the true degradation is a function of the observed degradation and the measurement error. As the true degradation is not directly observed it is proposed to obtain an estimate based on a deconvolution operation, which considers the subtraction of random variables such as the observed degradation and the measurement error. Given that the true degradation is free of measurement error, the first-passage time distribution will be different from the observed degradation. For the establishment of a control mechanism, these two distributions are compared using different indices, which account to describe the differences between the observed and true degradation. By defining critical levels of these indices, the reliability assessment may be obtained under a known level of measurement error. An illustrative example based on a fatigue-crack growth dataset is presented to illustrate the applicability of the proposed scheme, the reliability assessment is developed, and some important insights are provided.

Keywords: deconvolution; gamma process; lifetime; measurement system analysis; reliability estimation

Citation: Rodríguez-Picón, L.A.; Méndez-González, L.C.; Romero-López, R.; Pérez-olguín, I.J.C.; Rodríguez-Borbón, I.; Valles-Rosales, D.J. Reliability Analysis Based on a Gamma-Gaussian Deconvolution Degradation Modeling with Measurement Error. *Appl. Sci.* **2021**, *11*, 4133. https://doi.org/10.3390/app11094133

Academic Editor: Cher Ming Tan

Received: 18 March 2021
Accepted: 27 April 2021
Published: 30 April 2021

Publisher's Note: MDPI stays neutral with regard to jurisdictional claims in published maps and institutional affiliations.

Copyright: © 2021 by the authors. Licensee MDPI, Basel, Switzerland. This article is an open access article distributed under the terms and conditions of the Creative Commons Attribution (CC BY) license (https://creativecommons.org/licenses/by/4.0/).

1. Introduction

Generally, the observed degradation of a performance characteristic of interest is an additive function of the true degradation, and some measurement error [1–3]. This means that in most cases, it is difficult to measure the degradation process over time due to imperfect measurement devices and environmental conditions. If the measurement system accuracy can be attained during the measuring process, then the general reliability assessment of the product under study may be deemed as precise. Nevertheless, in the presence of measurement error, the estimation and reliability assessment must consider the measurement error in the modeling such that the obtained conclusions may not be underestimated.

Several models proposed in the literature consider the problem of obtaining the true degradation in the presence of measurement error with the common assumption is that the measurement error is independent of the degradation measurement [2,4,5]; given that the error comes from a measuring device that is independent of the true degradation. However, in some cases, it is considered that the measurement error is dependent on the true degradation [6–9]. In addition, a common assumption is that the measurement error is normally distributed with mean zero and standard deviation σ [5,10,11]. The true degradation can be obtained by considering the joint distribution of the probability density function (PDF) of the observed degradation and the PDF of the measurement error as described in the works of Pulcini [8], Lu et al. [7], Xie et al. [12], Kallen and van Noortwijk [13]. In these cases, the joint distribution is obtained either via joint conditional distributions or by the convolution of the observed degradation and the error [9]. In terms of stochastic modeling, the Wiener process is the most used in the literature to deal with measurement error. Shen et al. [14] proposed a Wiener process model with logistic distributed measurement errors, the estimation of parameters was carried out with the Monte Carlo Expectation–Maximization method to estimate the related parameters. Wang et al. [15] proposed a change-point Wiener degradation model with normally distributed measurement errors, they considered a Bayesian approach to estimate the parameters of interest. Pan et al. [16] studied a Wiener degradation model with three sources of uncertainty, one being the measurement error, which is considered to be normally distributed. Sun et al. [17] proposed a nonlinear Wiener process model with measurement error to estimate the remaining useful life of a cutting tool. The estimation of parameters of this model is extended by Tang et al. [18]. Liu and Wang [19] also considered the Wiener process with measurement error but based on evidential variables. Li et al. [20] proposed a Wiener process model with normally distributed measurement errors and multiple accelerating variables. Models based on the inverse Gaussian process with measurement error have also been proposed. Sun et al. [21], Chen et al. [22] and Hao et al. [23] studied the inverse Gaussian process with random effects and measurement errors to obtain lifetime estimations. A similar method for the inverse Gaussian process was also considered by [24] but under accelerating conditions. Chen et al. [25] proposed a nonlinear adaptive inverse Gaussian process with measurement error to estimate remaining useful life. Another important modeling approach considers the deconvolution, which consists of the inverse process of the convolution in order to obtain an unknown PDF from two known PDFs. In such a case, the true degradation can be obtained by deconvoluting the PDF of the observed degradation and the known PDF of the measurement error. Although, the deconvolution has been used in different scientific disciplines leading to important applications such as in illumina BeadArrays [12,26], optical distortion [27] and image processing [28,29]. It has only be considered to model the measurement error in degradation processes based on the inverse Gaussian and Wiener processes [30]. Furthermore, Rodriguez-Picon et al. [30] demonstrated the applicability of deconvolution to obtain reliability assessments without measurement error, but a control scheme over the performance of the measurement system is not considered. Important information about the deconvolution process can be found in Zinde-Walsh [31], Wang and Wang [32] and Neumann [33].

The importance of considering the measurement error in the modeling of degradation processes relies on obtaining accurate reliability assessments. However, it is also important to establish a control scheme over the measurement error, such that a desired estimation can be obtained under a controlled level of error. This means that a certain range of the observed measurement error caused by measuring devices, methods and environmental conditions can be established and maintained in order to achieve a desirable reliability assessment [4]. Usually, the reliability assessment based on degradation modeling is carried out by considering the first-passage time distribution of the degradation paths [34]. Thus, some variation of the first-passage times is expected from the observed degradation and the true degradation. This means that a certain level of permissible measurement error leads to a certain range of variation of the parameters of the first-passage time distribution. By controlling such variations, it is possible to obtain accurate life estimations, which are quite important in the definition of maintenance programs [35] or in the establishment of product warranties. Other approaches for reliability monitoring are based on control charts [36], these procedures are important to study the deterioration of systems which leads to determine maintenance policies and process availability improvement [37–39], as discussed in the degradation modeling with measurement error.

The main focus of this article is to establish a scheme to control the measurement error to obtain certain reliability assessments under a defined performance of the measurement system, where the performance is defined by the measurement error. The proposed scheme consists of first estimating the observed degradation parameters with measurement error under the gamma process. Then, the measurement system is assessed via a repeatability and reproducibiity (R&R) study in the aims of obtaining information about the total variance contribution of the measuring process. It is considered that the measurement error is normally distributed with mean zero and that an estimation of the standard deviation σ can be obtained from the total variation of the measurement system captured by the R&R study. The deconvolution approach is then performed to obtain the true degradation. As the function of the true degradation does not have a close analytical expression, we fitted the deconvoluted true degradation to different stochastic processes. Once the best fitting stochastic process is selected, the true first-passage distribution is characterized and compared to the observed first-passage time distribution by using different indices. Such indices are based on the coefficient of variation, the variance, the mean and the percentiles of the two distributions. By considering a critical value for any of the four indices, a certain level of performance of the coefficient of variation, variance, mean and percentile can be achieved under a certain value of σ. The defined value of σ can be used to control the measurement system in order to obtain a desired accuracy of the reliability assessment. The proposed scheme is implemented in a case study which consists of crack propagation data of an electronic device.

The rest of the article is organized as follows. In Section 2, the modeling of the observed degradation based on a gamma process is presented. In Section 3, the method to obtain the true degradation based on deconvolution is introduced. In Section 4, the proposed indices to compare the first-passage time distributions of the observed and true degradation are presented. In Section 5, a case study based on the crack propagation data of a electronic device is presented, the proposed scheme is implemented and the reliability assessment is developed under a defined level of measurement error. In Section 6, an extension for logistic distributed measurement errors is presented and illustrated. Finally, in Section 7, the discussion and some concluding remarks are provided.

2. Modeling of the Observed Degradation via Gamma Process

In this article, it is considered that the degradation measurements of a certain performance characteristic are contaminated with measurement error. These measurements are considered as the observed degradation, as these are directly observed. The modeling of this characteristic is firstly discussed in this section. In this case, stochastic modeling of the degradation process is considered given that it is possible to introduce the temporal uncertainty of the degradation increments over time [40]. The gamma process is specifically considered to describe the observed degradation of a characteristic of interest. This process has been widely documented and implemented in multiple case studies in the literature [40–43], this given its characteristics that it is a monotone stochastic process with independent and non-negative increments.

We consider $\{Z(t), t \geq 0\}$ as a degradation process that describes the observed degradation of a performance characteristic over time, it is deemed that $Z(t)$ is governed by a gamma process with the following properties: the degradation increments $Z(t + \Delta t) - Z(t) = \Delta Z(t)$ follow a gamma distribution $Ga(v[t + \Delta t - t], u)$, and $\Delta Z(t)$ are independent $\forall t_1 < t_2 < t_3 < t_4$.

From the PDF of the gamma process, the parameter vt is a non-negative shape parameter with $t \geq 0$, $v(0) \equiv 0$, while $u > 0$ is the scale parameter. It is known that the mean and variance of the processes are defined as $vt \cdot u$ and $vt \cdot u^2$, respectively. Thus, $\Delta Z(t)$ has the following PDF,

$$f(\Delta Z(t)|v,u) = \frac{\Delta Z(t)^{vt-1}}{u^{vt}\Gamma(vt)} exp\left\{-\frac{\Delta Z(t)}{u}\right\}. \qquad (1)$$

An important aspect of the reliability assessment of degradation processes is related to the first-passage times, these are events described by the moment when the cumulative degradation reaches a critical level ω. Thus, the first-passage time of the observed degradation is defined as $T_o = inf\{t_o : Z(t) \geq \omega\}$. The cumulative degradation can be used to describe the cumulative distribution function (CDF) of the first-passage times as $P(Z(t) \geq \omega) = 1 - F_{Ga}(\omega, v, u)$, which results as,

$$P(Z(t) \geq \omega) = \int_\omega^\infty f_{Z(t)}(z)dz = \frac{\Gamma(vt, \omega/u)}{\Gamma(vt)}. \quad (2)$$

The first-passage time CDF in (2) can be related to the Birnbaum–Saunders distribution [44,45], with parameters $\alpha_o^* = \sqrt{u/(\omega - z_0)}$ and $\beta_o^* = (\omega - z_0)/uv$, where z_0 is the initial level of degradation. The CDF is defined as follows

$$F_{T_o}(t) = \Phi\left[\frac{1}{\alpha_o^*}\left(\sqrt{\frac{t}{\beta_o^*}} - \sqrt{\frac{\beta_o^*}{t}}\right)\right], \quad (3)$$

where Φ denotes the standard normal CDF. The mean of the first-passage time distribution is obtained as $E(T_o) = \beta_o^*(1 + \alpha_o^{*2}/2)$, and the variance as $Var(T_o) = \alpha_o^*\beta_o^*(1 + 5\alpha_o^{*2}/4)^{1/2}$. As (1) denotes the PDF of the observed degradation, let us consider a scheme of a degradation test where $i = 1, 2, \ldots, N$ units are tested and $j = 1, 2, \ldots, M$ denotes the total number of measurements for all the tested units, which results in observed degradation measurements $Z_i(t_j)$. Then, it is defined that the degradation increments $\Delta Z_i(t_j) = Z_i(t_j) - Z_i(t_{(j-1)})$, with $t_0 = 0$, $\Delta t_j = t_j - t_{(j-1)}$, have the next PDF,

$$f(\Delta Z_i(t_j)|v, u) = \frac{\Delta Z_i(t_j)^{v\Delta t_j - 1}}{u^{v\Delta t_j}\Gamma(v\Delta t_j)}exp\left\{-\frac{\Delta Z_i(t_j)}{u}\right\}. \quad (4)$$

As mentioned earlier, normally the observed degradation is contaminated with measurement error. Which implies that the true degradation cannot be observed directly from the degradation process. In such a case, it is important to find the true degradation in terms of the observed degradation PDF and an assumed PDF of the error.

3. Obtaining the True Degradation Distribution via Deconvolution

In this section, it is considered that $Z_i(t_j)$ represents the observed degradation measurement of the ith unit at time t_j, and that the observed degradation is contaminated with some measurement error ε_{ij}. Thus, ε_{ij} is also observed at t_j for each ith unit. Which means that for each observed degradation a measurement error is observed. Such that ε_{ij} is a random variable that follows a Gaussian distribution as $G(\mu, \sigma)$ with a PDF defined as,

$$f(\varepsilon_{ij}|\mu, \sigma) = \frac{1}{\sqrt{2\pi}\sigma}exp\left\{-\frac{(\varepsilon_{ij} - \mu)^2}{2\sigma^2}\right\}. \quad (5)$$

Based on this measurement error, an additive function of the observed degradation can be considered as $Z_i(t_j) = S_i(t_j) + \varepsilon_{ij}$, where $S_i(t_j)$ denotes a hidden true degradation measurement. Indeed, the observed degradation and measurement error are considered to be known as (4) and (5), respectively. Then, the true degradation may be obtained via deconvolution [30]. This operation consists in obtaining the subtraction of random variables, for example consider the function $H = E + G$, where H represents an observed measurement, E represents an unknown variable and G represents a measurement error. The PDFs of H and G are known to be f_H and f_G, respectively, and the characteristic functions (CF) of such PDFs are defined as φ_H and φ_G. The CFs are also known as Fourier transforms (FT). In the first instance, the deconvolution operation consists of determining the CF of E, which is defined as $\varphi_E = \varphi_H/\varphi_G$. In the second instance, the function of the deconvoluted true measurement f_E is obtained by considering the inverse Fourier transform (IFT) of φ_E.

Consider this approach for $Z_i(t_j) = S_i(t_j) + \varepsilon_{ij}$, where a gamma distribution is defined for $Z_i(t_j)$ as $f(Z_i(t_j))$ and a Gaussian distribution is defined for ε_{ij} as $f(\varepsilon_{ij})$. Firstly, the CF of $S_i(t_j)$ can be obtained by considering the CF of the gamma and normal distributions in (6) and (7), respectively.

$$\varphi_Z(\zeta) = (1 - ui\zeta)^{-v}. \tag{6}$$

$$\varphi_\varepsilon(\zeta) = exp\{i\mu\zeta - \sigma^2\zeta^2/2\}. \tag{7}$$

Thus, $\varphi_S(t)$ is obtained as

$$\varphi_S(\zeta) = \frac{(1 - ui\zeta)^{-v}}{exp\{i\mu\zeta - \sigma^2\zeta^2/2\}}. \tag{8}$$

The PDF of $S_i(t_j)$ is obtained via the IFT of (8) as,

$$f(S_i(t_j)) = \int_{-\infty}^{\infty} \varphi_S(\zeta) exp\{-i\zeta s\} d\zeta. \tag{9}$$

It can be noted that the IFT represented by the integral in (9) does not have a closed analytical expression. For this, the discrete Fourier transform (DFT) is considered to obtain an approximation of $f(S_i(t_j))$. The DFT considers a discrete version of the IFT in (9) as a Riemann sum approximation which can solved via the fast Fourier transform (FFT) algorithm [46,47]. The FFT is known to reduce the complexity of the DFT of a function sampled on a regular grid of 2^p points [48,49]. By considering that any integral, such as (9), can be viewed as the sum of infinitely many small rectangles, then for the sampled regular grid, p equally spaced sub-intervals with range $[-L_0, L_0]$ are considered, where $L_0 = \mu + 5\sigma + q_g$, q_g is the 0.99999 quantile of the gamma distribution, $-L_0 = 0$ and (μ, σ) are the parameters of the measurement error PDF. Both limits of the regular grid are defined considering the domain of the deconvoluted random variable, such that the minimum value of the deconvoluted observation is 0, and the maximum value corresponds to a L_0 as defined. Thus, the approximation of (9) can be viewed as the Riemann sum approximation [50] of the continuous IFT, as follows,

$$\simeq \frac{2L_0}{P} \sum_{j=0}^{P-1} \varphi_S\left(\frac{2L_0}{P}(j-1) - L_0\right) exp\left\{-i\zeta s\left(\frac{2L_0}{P}(j-1)\right) - L_0\right\}, \tag{10}$$

where, the width of the p equally spaced sub-intervals is defined as $2L_0/P$. The number of sub-intervals is considered to be a large enough integer number to obtain a good approximation of $f(S_i(t_j))$. The "NormalGamma" package [51] from R is used to implement the DFT in (10). As this package is defined for convolution operations, the original code was modified to implement the deconvolution operation.

Fortunately, the FFT algorithm can be implemented in R to solve the proposed DFT. Specifically, the function is defined as follows for the sampled vector $k = 0, \ldots, P-1$ [26],

$$W[k] = \frac{L_0}{P\pi} exp\{i\pi(k-1)\} \sum_{j=0}^{P-1} \varphi_S\left(\frac{2L_0}{P}(j-1) - L_0\right) exp\left\{\frac{2i\pi}{P}(j-1)(k-1)\right\}. \tag{11}$$

In this paper, $W[k]$ is considered to be an approximation of the true degradation, such that $W[k]$ is governed by a certain stochastic process. The gamma process, inverse Gaussian (IG) process, geometric Brownian motion (GBM) process and the Wiener process may be considered and the best fitting model may be selected by assessing their respective goodness of fit.

It should be noted that both the gamma and the IG processes are monotone processes, while the Wiener process is known to be non-monotone. If the observed degradation paths are monotone, then it is expected that the true degradation paths remain monotone. Which, can only be true when σ is small enough to sustain that $\Delta Z_i(t_j) > \epsilon_{ij}$. If σ is large enough such that $\Delta Z_i(t_j) < \epsilon_{ij}$, then the degradation paths may become non-monotone, which in some case studies may not be reasonable (such as in crack propagation data). Given that in this paper it is considered that the observed degradation paths are governed by a gamma process, it is expected that the true degradation remains governed by a monotone stochastic process. In Figure 1, a comparison of observed degradation paths and deconvoluted true degradation paths is presented when $\Delta Z_i(t_j) > \epsilon_{ij}$ and $\Delta Z_i(t_j) < \epsilon_{ij}$. The paths with black lines were simulated from a gamma process. From Figure 1a, it can be noted that if σ is large enough the true deconvoluted degradation paths become non-monotone. While in Figure 1b, it can be noted that if σ is small enough, the deconvoluted paths remain monotone.

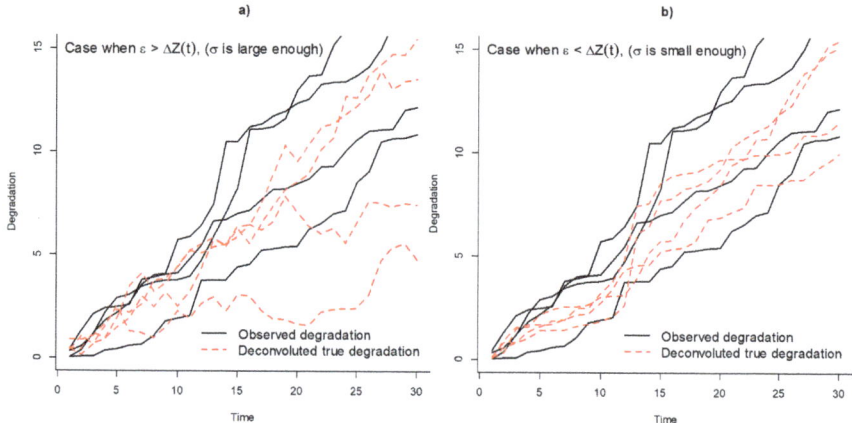

Figure 1. Comparison of degradation paths for observed and true degradation. (**a**) True non-monotone degradation paths in red dotted lines when $\Delta Z_i(t_j) < \epsilon_{ij}$, (**b**) true monotone degradation paths in red dotted lines when $\Delta Z_i(t_j) > \epsilon_{ij}$.

The construction of the deconvoluted paths (red dashed lines in Figure 1) is performed considering that the deconvolution operation is performed at every t_j for every degradation measurement $Z_i(t_j)$. Then, random true measurements of $S_i(t_j)$ are generated at every t_j to construct the different paths, which represents cumulative sums of the generated random variables. Once the best fitting stochastic process of $W[k]$ is defined, the true first-passage time distribution can be obtained. The lifetime of the true degradation is defined as $T_s = inf\{t_s : S_i(t_j) \geq \omega\}$. The first-passage time distribution will depend on the best fitting stochastic process.

4. The Effect of the Measurement Error over the First-Passage Time Distributions

It is expected that the measurement error affects the behavior of the first-passage time distributions of the observed degradation and the true degradation. If the measurement error is not considered in the modeling, the reliability assessment may be underestimated. For these reasons, it is important to study the effect of the measurement error over the first-passage time distributions, such that a maximum level of error in the measurement system can be determined to obtain a desired reliability assessment. The analysis in this section is focused on determining the differences between the PDF of T_o and T_s. Si et al. [4] proposed to compare two first-passage time distributions via the coefficient of variation (CV) and the variation ($Var(T)$) of two distributions. They implemented such approach in a Wiener model with measurement error. The CV can describe the amount of variability in any random variable, thus it is expected that the difference between the CV of T_o and T_s is relatively small if the effect of measurement error is small. The same approach can be considered if the corresponding variations $Var(T)$ and means $E(T)$ are compared. In this article, the indices proposed by Si et al. [4] are considered and described in (12) and (13) for the CV and variances, respectively. In addition, an index considering the means is proposed in (14). In fact, any percentile (z_q) of interest of the first-passage time distributions can be compared as described in (15).

$$I_{CV}(T_s, T_o) = \frac{|CV(T_s) - CV(T_o)|}{CV(T_o)}. \tag{12}$$

$$I_{Var}(T_s, T_o) = \frac{|Var(T_s) - Var(T_o)|}{Var(T_o)}. \tag{13}$$

$$I_E(T_s, T_o) = \frac{|E(T_s) - E(T_o)|}{E(T_o)}. \tag{14}$$

$$I_{z_q}(T_s, T_o) = \frac{|z_q(T_s) - z_q(T_o)|}{z_q(T_o)}. \tag{15}$$

The four indices are considered to describe the differences between the first-passage time distributions. In this way, it is expected that the four indices do not exceed critical values $\left(C_{CV}, C_{var}, C_E, C_{z_q}\right)$, which means that the estimated lifetime obtained from the distribution $F_{T_o}(t)$ can approach to the estimation of the distribution $F_{T_s}(t)$ under certain permissible level of measurement error described in the four critical levels $\left(C_{CV}, C_{var}, C_E, C_{z_q}\right)$. The flow chart in Figure 2 is followed to optimize σ in order to establish a control over the measurement system for a certain accuracy of the reliability assessment of interest.

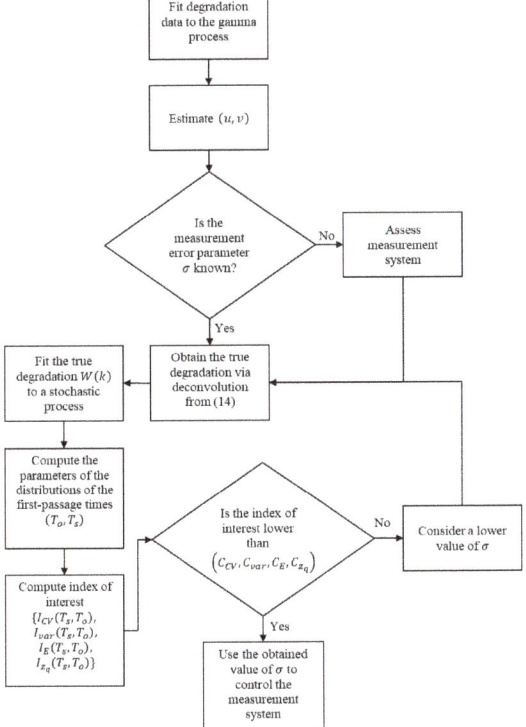

Figure 2. Proposed scheme for the optimal reliability analysis of degradation processes with measurement error.

5. Case Study

The dataset presented by Rodríguez-Picón et al. [52] is considered for the implementation of the proposed approach. This case study consists in the crack-growth of a terminal in an electronic device. The function of this terminal is to transfer a signal to a receptor, which can be disrupted if the crack in the terminal propagates to a certain critical level, and thus would lead to a failure of the device. A DT was carried out to study the propagation of the crack in 10 terminals. The crack propagation was measured every 0.1 hundred thousand cycles until 0.9 hundred thousand cycles. In this article, it is considered that a failure is said to have occurred when the length of the crack exceeds the critical length of 0.4 mm. The total of sample devices are $N = 10$, with $M = 9$ observation times as $j = 1, 2, 3, 4, 5, 6, 7, 8, 9$, which are the same for all the $i = 1, 2, \ldots, 10$ samples with $t_j = (0.1, 0.2, 0.3, 0.4, 0.5, 0.6, 0.7, 0.8, 0.9)$ hundred thousand cycles. In Table 1, the degradation measurements are presented, the units are millimeters. In Figure 3, the cumulative degradation paths are presented.

Table 1. Degradation dataset of crack-growth case study.

Device	Hundred Thousands of Cycles									
	0	0.1	0.2	0.3	0.4	0.5	0.6	0.7	0.8	0.9
1	0	0.01	0.03	0.055	0.107	0.165	0.183	0.2	0.26	0.302
2	0	0.09	0.161	0.172	0.247	0.259	0.281	0.371	0.401	0.429
3	0	0.01	0.06	0.081	0.118	0.142	0.158	0.169	0.232	0.262
4	0	0.016	0.076	0.087	0.104	0.127	0.198	0.208	0.218	0.258
5	0	0.036	0.096	0.176	0.204	0.242	0.281	0.325	0.415	0.495
6	0	0.014	0.102	0.112	0.194	0.277	0.289	0.305	0.335	0.391
7	0	0.037	0.064	0.078	0.096	0.124	0.164	0.234	0.254	0.326
8	0	0.035	0.086	0.105	0.174	0.267	0.277	0.347	0.361	0.384
9	0	0.067	0.148	0.161	0.173	0.184	0.218	0.229	0.239	0.285
10	0	0.025	0.052	0.064	0.076	0.151	0.187	0.205	0.222	0.262

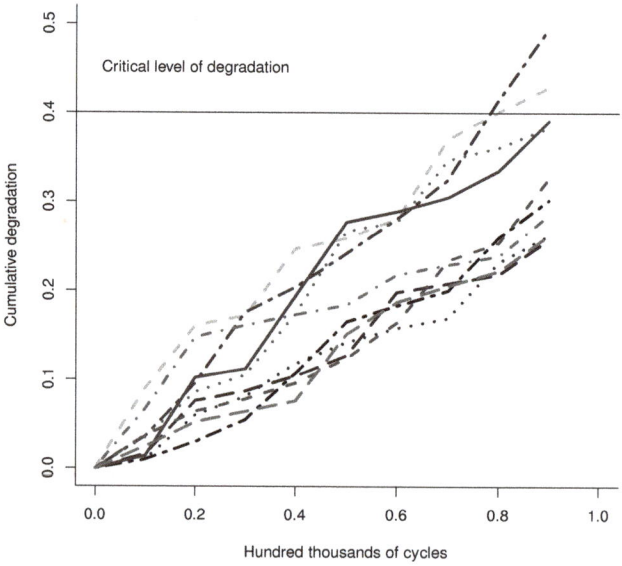

Figure 3. Cumulative degradation paths of the case study.

It is assumed that the degradation data in Table 1 are governed by a gamma process as in (4). Thus, $Z_i(t_j)$ is the observed degradation for $i = 1, 2, \ldots, 10$, and

$t_j = (0.1, 0.2, 0.3, 0.4, 0.5, 0.6, 0.7, 0.8, 0.9)$. In this case study, it is considered that the measurement error is described by a Gaussian distribution with $\mu = 0$ and σ, as described in (5). Thus, by considering the flow chart in Figure 2, we first estimate the parameters of the gamma process for the degradation dataset in Table 1, then we estimate σ by performing an R&R study to the measurement system. Next, we illustrate the effect of the measurement error over the true degradation distribution by using the deconvolution modeling proposed in (9), and to assess the effect over the free-error first-passage time distribution by using the indices presented in ((12)–(15)).

5.1. Estimation of Parameters for the Observed Degradation

The parameters of the observed degradation (u, v) are estimated via Bayesian approach by considering informative gamma prior distributions for the unknown parameters (u, v), as $v \sim Ga(\zeta, \eta)$, $u \sim Ga(\delta, \tau)$. Where, the shape parameters are $\zeta = 52.79$ and $\delta = 41.45$, and the scale parameters are $\eta = 0.4257$ and $\tau = 4.13 \times 10^{-4}$, for v and u, respectively. The Markov chain Monte Carlo (MCMC) algorithm is utilized to sample from the joint distribution based on the Gibbs sampler. For this, a code is developed in the software OpenBUGS [53]. A total of 50,000 iterations were considered for burn-in purposes and 100,000 iterations were considered for estimations purposes. The obtained estimations for the mean, standard deviation, Monte Carlo error, and some percentiles for the parameters (u, v) are presented in Table 2. Two sets of initial values are considered in order to assess the convergence of the parameters with the Brooks–Gelman–Rubin (BGR) statistic, the obtained graphs from OpenBUGS are presented in Figure 4. It is considered that convergence is achieved if all the lines in Figure 4 transpose in 1 [54]. It can be noted that convergence is achieved in both parameters.

Table 2. Obtained estimations for the observed degradation.

Parameter	Mean	Sd	MC Error	$p_{0.025}$	$p_{0.5}$	$p_{0.975}$
v	22.55	3.094	0.01476	16.93	22.39	29.02
u	0.01664	0.002687	1.76×10^{-5}	0.0126	0.01658	0.02309

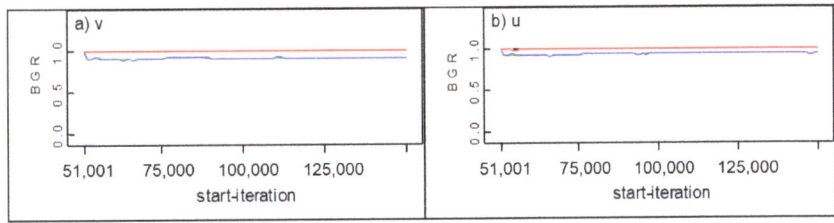

Figure 4. BGR graphs for parameters of the observed degradation gamma process, (a) v, (b) u.

The first-passage time distribution of the observed degradation is obtained from (3). By considering the mean estimates from Table 2, and $z_0 = 0$, the parameters can be obtained from $\alpha_o^* = \sqrt{u/(\omega - z_0)}$ and $\beta_o^* = (\omega - z_0)/uv$ as $\hat{\alpha}_o^* = 0.2039$ and $\hat{\beta}_o^* = 1.066$. With these estimates, it is easy to compute the mean and variance as $E(T_o) = 1.088$ and $Var(T_o) = 0.223$, thus $CV(T_o) = 0.434$.

5.2. Characterization of the Measurement Error and Its Effect

The degradation increments in Table 1 were measured using a vision system with special software applications to measure crack propagations. As σ is unknown, we performed an R&R study to assess the performance of the measurement system and to determine how much of the observed variation is due to the measurement system variation, i.e., σ. The study was performed under the next characteristics: a total of three people were selected to perform the study, 10 devices were selected, and three replicates were performed, making a total of 60 readings. The results of the gage R&R study are presented in Table 3 and Figure 5. It can be noted from Table 3 that the total variation contribution of the repeatability and reproducibility are 3.83% and 0.00%, respectively, which makes the total gage R&R contribution at 3.83%. The general rule says that if the total gage R&R contribution is less than 10%, the measurement system is acceptable, which is the case of this study. From Figure 5, it can be noted that indeed the measurement system performs well, and that most of the variation comes from the part-to-part variation.

Table 3. Gage R&R variation contribution.

Source	StdDev (SD)	Study Variation (6*SD)	% Study Variation
Total gage R&R	0.0006058	0.0036347	3.83
Repeatability	0.0006058	0.0036347	3.83
Reproducibility	0.0000000	0.0000000	0.00
Operators	0.0000000	0.0000000	0.00
Part to Part	0.0157952	0.0947713	99.93
Total Variation	0.0158068	0.0948409	100

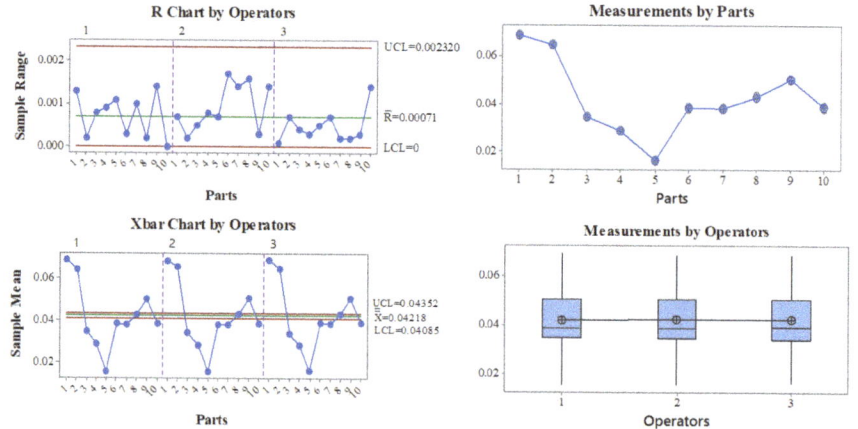

Figure 5. Graphs obtained from the gage R&R study applied to the measurement system.

From Table 3, the standard deviation of the gage R&R study is $\sigma_{R\&R} = 0.0006058$, which is the total variation due to the measurement system. Thus, we consider $\sigma_{R\&R}$ as an estimation of σ, such that $\hat{\sigma} = \sigma_{R\&R} = 0.0006058$, and use this value to perform the deconvolution approach presented in Section 3.

Considering the estimated parameters in Table 2 for the gamma process, we estimated q_g for every t_j. Then, we implemented the deconvolution approach considering $\hat{\sigma} = 0.0006058$ and $p = 1000$. In Figure 6, a comparison of the observed degradation paths and the obtained true deconvoluted degradation paths is provided by presenting the box plots and mean for every t_j. It can be noted from Figure 6 that the variation in every t_j was reduced in the true deconvoluted paths. In addition, the mean degradation in every t_j is smaller in the true deconvoluted paths than the observed paths as in the work of Rodriguez-Picon et al. [30]. It is obvious, that the reduction in the variation at every t_j will cause variations in the mean degradation, i.e., degradation rate, as can be noted in both degradation paths. Indeed, these conditions will have an impact on the first-passage time distributions.

As the true degradation function does not have an analytical closed form, we consider to fit the obtained true degradation to the gamma, IG, GBM and Wiener stochastic processes. Stochastic models are considered to describe temporal uncertainty, such as the observed degradation is modeled with the gamma process. Although, a simple approximation can be defined when obtaining cumulative sums of the true deconvoluted variables. To perform a reliability assessment of such approximation, the Kaplan–Meier method can be considered to define the reliability function. In order to assess the goodness of fit of the four stochastic processes we consider a graphical method such as the Q-Q plots. In Figure 7, the Q-Q plots for the different stochastic processes are presented for the true degradation.

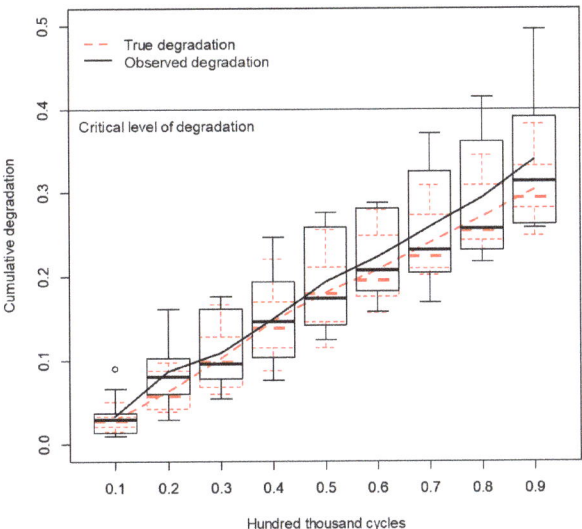

Figure 6. Illustration of differences between the observed degradation paths and true degradation paths.

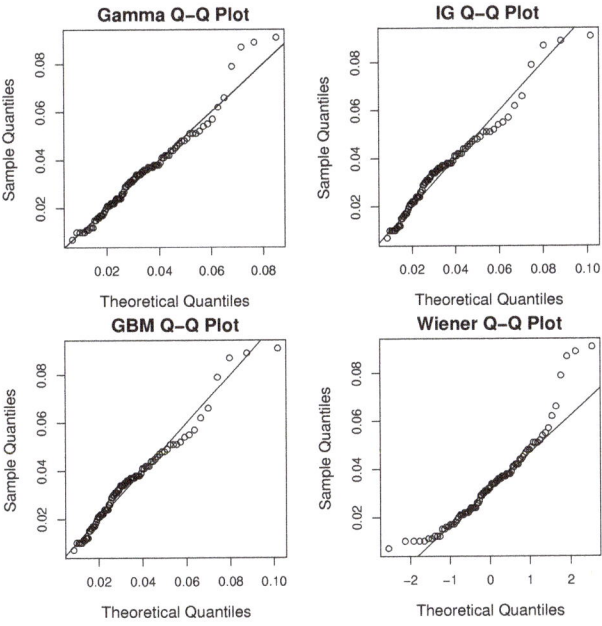

Figure 7. Q-Q plots for true degradation under different stochastic processes.

It can be noted from Figure 7 that the gamma process seems to have a better fit. In addition to the Q-Q plots, we also performed the Cramér–von Mises goodness of fit test for all the models. The obtained Cramér–von Mises statistics were, for gamma 0.056918, for IG 0.20057, for GBM 0.1748 and for Wiener 0.18345. By considering the critical value for the Cramér–von Mises statistic for a significance level of 0.1 of 0.173, it can be noted that the gamma process is the only one not rejected. Thus, we consider the gamma process to govern the true deconvoluted degradation.

5.3. Comparison of the First-Passage Time Distributions

It is considered that the vector $W[k]$ is described by a gamma process as $Ga(v^*\Delta t_j, u^*)$ with shape parameter $(v^*\Delta t_j)$, and scale parameter (u^*). The true first-passage time distribution can be obtained by considering the Birnbaum–Saunders distribution with parameters $\alpha_s^* = \sqrt{u^*/(\omega - z_0)}$ and $\beta_s^* = (\omega - z_0)/v^*u^*$, with $z_0 = 0$, and $\omega = 0.4$. Thus, the CDF is described as

$$F_{T_s}(t) = \Phi\left[\frac{1}{\alpha_s^*}\left(\sqrt{\frac{t}{\beta_s^*}} - \sqrt{\frac{\beta_s^*}{t}}\right)\right], \tag{16}$$

with mean obtained as $E(T_s) = \beta_s^*(1 + \alpha_s^{*2}/2)$, and the variance as $Var(T_s) = \alpha_s^*\beta_s^*(1 + 5\alpha_s^{*2}/4)^{1/2}$.

The estimated gamma parameters of the true degradation were obtained as $\hat{v}^* = 37.9237$ and $\hat{u}^* = 0.0088$. Considering these estimates, the parameters of the first-passage time distributions for the true degradation can be easily obtained considering the Birnbaum–Saunders distribution. The computed parameters were obtained as $\hat{\alpha}_s^* = 0.1483$ and $\hat{\beta}_s^* = 1.1985$. In Table 4, a comparison of the the mean, variance, and CV for the observed and true first-passage time distributions is presented.

Table 4. Comparison of the mean, variance and CV for the first-passage times of the observed and true degradation.

	Mean	Variance	CV
Observed	1.088	0.223	0.434
True	1.211	0.18	0.1487

From Table 4, the effect of the measurement over the first-passage time distribution is reflected. For instance, the mean passage-time from the true degradation is greater that the obtained from the observed degradation. In addition, the variance is smaller for the true degradation compared to the observed degradation. This finding can be confirmed by the degradation paths described in Figure 6, where the mean degradation and the variation among degradation paths are smaller compared to the observed degradation.

The distributions $f_{T_o}(t)$ and $f_{T_s}(t)$ are compared by computing the indices described in ((12)–(15)) as denoted in the flow chart in Figure 2. The 5th percentile is considered for (15). The quantile function of the Birnbaum–Saunders distribution is described as [55],

$$t(q) = \beta\left(\frac{\alpha z_q}{2} + \sqrt{\frac{\alpha^2 z_q^2}{4} + 1}\right)^2,$$

where z_q is the $q \times 100$th quantile of the standard normal distribution. Considering that $z_5 = -1.6448$ and the estimates $\hat{\alpha}_o^*, \hat{\beta}_o^*$ and $\hat{\alpha}_s^*, \hat{\beta}_s^*$, the 5th percentile for the first-passage time distributions for the observed and true degradation were obtained as $t_o(5) = 0.7633$ and $t_s(5) = 0.9396$, respectively. All four indices described in Section 4, $I_{CV}(T_s, T_o)$, $I_{Var}(T_s, T_o)$, $I_E(T_s, T_o)$, $I_{z_5}(T_s, T_o)$ were computed and are presented in Table 5.

Table 5. Computed indices from the observed and true first-passage time distributions.

	$I_{CV}(T_s, T_o)$	$I_{Var}(T_s, T_o)$	$I_E(T_s, T_o)$	$I_{z_5}(T_s, T_o)$
Index	0.6574	0.1919	0.1136	0.2309

The critical values ($C_{CV}, C_{var}, C_E, C_{z_5}$) may be defined depending of the allowance for the measurement error of the measurement system. In this paper, four indices are considered, however, one index can be used depending of the reliability estimation of interest. As can be noted from ((12)–(15)), the indices are ratios that account for the relative increase in each reliability estimation, i.e., CV, variance, mean, percentiles. Indeed, the greater the value of the respective indices the more difference between reliability estimations, which means more variability of the measurement error, i.e., σ. Critical values should be defined based on historical behavior, but a first approach can be considered as follows: consider that the measurement system has been evaluated and an estimation of σ is obtained, then the true and observed first-passage time distributions can be characterized. Consider that σ is a component of variance with good performance, i.e., less than 10% of the total variation of the process [56]. If the mean is the estimation of interest, then from (14) it can be considered that $C_E = I_E$, this equivalence of the critical value will relate the good performance of the measurement system to the estimation of the mean failure time, so the control scheme can be initiated. Now consider a specific example, if the critical values are considered as $C_{CV} = 0.66$, $C_{Var} = 0.2$, $C_E = 0.12$, and $C_{(z_5)} = 0.24$, it is observed from Table 5 that the standard deviation for the measurement error $\sigma = 0.0006058$ is good enough for the performance of the measurement system. With this parameter of the measurement error, it is expected that the estimated lifetime from the data contaminated with measurement error should be accurate enough. It should be noted that σ may be optimized by following the sequence determined in the flow chart in Figure 2. However, previous knowledge of the case study must be available such that optimal values of the critical values ($C_{CV}, C_{var}, C_E, C_{z_1}$) are defined. For instance, the mean time to failure (MTTF) of the observed degradation and the true degradation with $\sigma = 0.0006058$ are $E(T_o) = 1.088$, and $E(T_s) = 1.211$, respectively. Which means a difference of 0.123 hundred thousands of cycles. If the maximum allowance for the difference between MTTF is expected to be, for example, 0.05 hundred thousands of cycles and the critical value $C_E = 0.043$, it can be noted that $I_E(T_s, T_o)$ is higher C_E which means that the measurement system should be improved such that the measurement process is executed more accurately and the variation caused by the measurement system is reduced. The same approach may be considered for any other index different than C_E, depending on the estimation of interest. For example, if it is expected that the differences between the 5th percentiles of the failure times for the observed degradation and the true degradation be 0.09 hundred thousands of cycles and $C_{z_5} = 0.105$, again it can be noted that $I_{z_5} > C_{z_5}$, which denotes a high variance of the measurement error.

The reliability functions with and without measurement error were obtained based on the corresponding first-passage time distributions. The respective differences can be noted in Figure 8. Along with the respective reliability functions, we also present the Kaplan–Meier reliability for the observed and true degradation, along with their respective 95% confidence intervals. At different t_j, the Kaplan–Meier confidence interval of the true reliability does not include the observed reliability, which denotes the difference. A difference of the reliability functions presented by Rodriguez-Picon et al. [30], apart from the considered stochastic processes for the observed degradation, relies on that, in this paper, a stochastic process is fitted to the true degradation which defines the dashed red reliability function. Furthermore, the reliability estimation from Kaplan–Meier results in a different behavior as the true degradation comes from a gamma-Gaussian deconvolution.

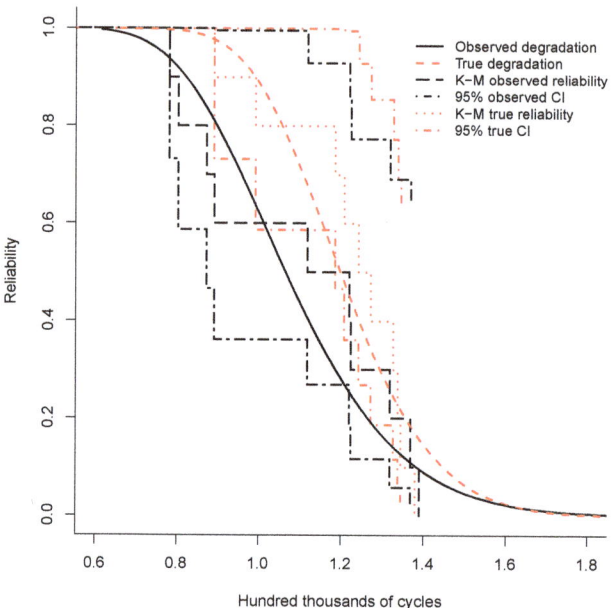

Figure 8. The effect of the measurement error illustrated by the comparison of the estimated reliability functions and the Kaplan–Meier estimation with confidence intervals.

6. Extension for Non-Gaussian Measurement Errors

Other PDFs can be considered to describe the measurement error in the proposed approach. As the deconvolution operation is performed based on CFs, the proposed method can be extended to more PDFs by replacing the corresponding CF in the denominator of (8). Then, the approximation of the true degradation can be obtained by implementing the fast Fourier transform in (10) and (11). In this section, we illustrate this extension by considering that the measurement error follows a logistic distribution $f_L(\varepsilon_{ij}|\mu_L, s)$. The CF is presented as follows,

$$\varphi_\varepsilon(\zeta) = exp\{i\zeta \mu_L\} \frac{\pi s \zeta}{sinh(\pi s \zeta)}, \tag{17}$$

Then, the CF of the true measurement is defined as,

$$\varphi_S(\zeta) = \frac{(1 - ui\zeta)^{-v}}{exp\{i\zeta \mu_L\} \frac{\pi s \zeta}{sinh(\pi s \zeta)}}, \tag{18}$$

The CF in (18) is considered in (11) to obtain the true measurements. The parameters of the observed degradation are presented in Table 2 as $v = 22.55$ and $u = 0.01664$. From the R&R study, it is known that the total variation due to the measurement system is $\hat{\sigma} = \sigma_{R\&R} = 0.0006058$. For this scenario, it is considered that $\mu_L = 0$, and as the standard deviation of the logistic distribution is defined as $SD = \sqrt{(s^2 \pi^2/3)}$, then by considering $SD = 0.0006058$ it follows that $s = 0.0033$. The deconvolution approach is implemented considering these parameters with $p = 1000$. The vector $W[k]$ was then fitted to the gamma, Wiener, inverse Gaussian and GBM processes. It was found that the gamma process is the best fitting model. The reliability function based on the estimated parameters of the first-passage time distributions with logistic errors is compared with the Gaussian errors in Figure 9. It can be noted that the behavior of the reliability functions is quite similar. With the logistic errors, the reliability is estimated to be greater when $t > 1.1$ hundred thousand cycles, approximately. It is known that the logistic distribution has higher kurtosis than the Gaussian distribution, which may account for the small differences in the reliability function. Both reliability functions, estimated considering measurement error, determine that the true degradation has greater reliability than the observed degradation.

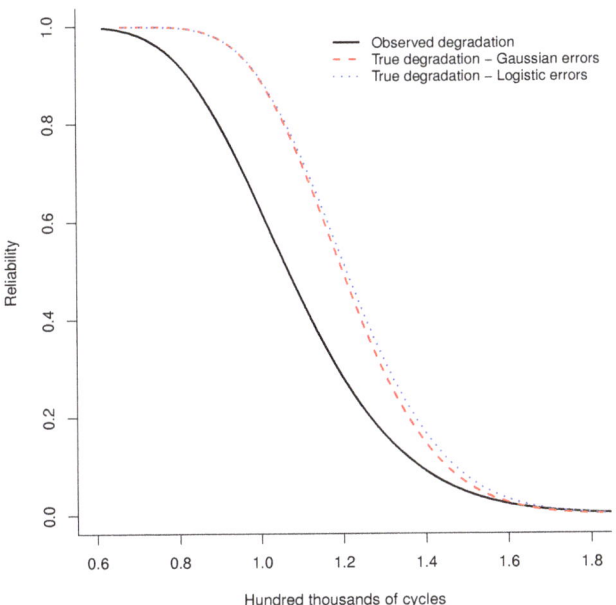

Figure 9. Comparison of reliability functions for the observed degradation and the true degradation under Gaussian and logistic measurement errors.

7. Concluding Remarks and Discussion

The reliability assessment of products is a critical activity for different processes and systems, thus it is important to consider that the analyzed data are free of any contamination that can cause inaccurate conclusions. Furthermore, as the measuring process is an integral part of reliability testing, it is also important to establish some control schemes over the measuring system's performance. Such that, a certain performance of the system leads to a predetermined performance of the product's reliability assessment. In the case of degradation modeling, the measuring error causes variation in the first-passage time distribution. Based on this, it may be expected that the reliability assessment under contaminated data may be underestimated. In this article, it is considered that a gamma process governs the observed degradation with measurement error, and it is assumed that the error can be described by a Gaussian distribution with mean zero and standard deviation σ. Thus, the true degradation is obtained by deconvoluting the observed degradation and the measurement error. In order to control the measurement error in terms of the reliability assessment, the first-passage time distributions of the observed and the true degradation are compared in terms of some proposed indices. A general scheme was proposed to establish the differences between distributions in order to obtain the desired accuracy of the assessment. From the case study, it was observed that depending on the reliability estimation of interest; it is possible to establish a maximum level of the standard deviation of the measurement error. This enables to control the measuring system. It is essential to define critical levels of the indices for the first-passage time distributions, such that a maximum level of error can be established. These critical values can be defined by considering the maximum difference between the reliability estimation of interest between the true and contaminated first-passage distributions. Following the proposed scheme, the permissible error can be determined as described in the case study. Furthermore, a scenario to deal with non-Gaussian measurement errors is presented to extend the deconvolution approach applicability.

There are several opportunities for further research in the proposed scheme of this article. Although the gamma process has been widely used in degradation modeling, other stochastic processes can be used to describe the observed degradation, such as the inverse-Gaussian process, geometric Brownian motion and the Wiener process. The deconvolution modeling proposed in this paper can be extended by considering any of these processes. Although, the implementation for some process may result more complex, as the CFs of the inverse Gaussian and geometric Brownian motion do not have closed expressions, which impose interesting challenges for the implementation of the deconvolution approach. Furthermore, other sources of uncertainty can be included in the degradation modeling. It has been found that the consideration of random effects accounts for the accuracy of the reliability estimations. Indeed, these sources imply certain mathematical complexity which should be added to the computational complexity of the deconvolution approach. For this, different deconvolution algorithms proposed in the literature may be considered to obtain approximations of true variables obtained from measurement error contaminated processes. The CV, variation, mean and percentiles are considered as indices to measure the differences between first-passage time distributions. Nevertheless, some other metrics can be studied with the same purpose. In addition, we consider some well-known stochastic processes to model the true degradation as an approximation, given that the function of the true degradation does not have a closed analytical expression. However, further investigation may be directed in the future to study the deconvoluted function of gamma and Gaussian distributions.

Author Contributions: Conceptualization, L.A.R.-P.; methodology, L.A.R.-P. and L.C.M.-G.; software, L.A.R.-P. and L.C.M.-G.; validation, L.A.R.-P., I.J.C.P.-O. and M.I.R.-B.; formal analysis, L.A.R.-P.; investigation, L.A.R.-P. and L.C.M.-G.; resources, L.A.R.-P., I.J.C.P.-O., D.J.V.-R. and R.R.-L.; writing—original draft preparation, L.A.R.-P. and L.C.M.-G.; writing—review and editing, L.A.R.-P., L.C.M.-G., I.J.C.P.-O. and D.J.V.-R.; visualization, L.A.R.-P., R.R.-L and M.I.R.-B.; funding acquisition, R.R.-L., L.A.R.-P., I.J.C.P.-O., I.R.-B. and D.J.V.-R. All authors have read and agreed to the published version of the manuscript.

Funding: This research received no external funding, the APC was funded by the Autonomous University of Ciudad Juarez.

Conflicts of Interest: The authors declare no conflicts of interest.

References

1. Wang, Z.; Li, J.; Zhang, Y.; Fu, H.; Liu, C. A novel wiener process model with measurement errors for degradation analysis. *Eksploatacja i Niezawodnosc* **2016**, *18*, 396. [CrossRef]
2. Ye, Z.S.; Wang, Y.; Tsui, K.L.; Pecht, M. Degradation data analysis using Wiener processes with measurement errors. *IEEE Trans. Reliab.* **2013**, *62*, 772–780. [CrossRef]
3. Li, J.; Wang, Z.; Liu, X.; Zhang, Y.; Fu, H.; Liu, C. A Wiener process model for accelerated degradation analysis considering measurement errors. *Microelectron. Reliab.* **2016**, *65*, 8–15. [CrossRef]
4. Si, X.S.; Chen, M.Y.; Wang, W.; Hu, C.H.; Zhou, D.H. Specifying measurement errors for required lifetime estimation performance. *Eur. J. Oper. Res.* **2013**, *231*, 631–644. [CrossRef]
5. Whitmore, G. Estimating degradation by a Wiener diffusion process subject to measurement error. *Lifetime Data Anal.* **1995**, *1*, 307–319. [CrossRef]
6. Rabinovich, S.G. *Evaluating Measurement Accuracy*; Springer: Berlin/Heidelberg, Germany, 2010. [CrossRef]
7. Lu, D.; Xie, W.; Pandey, M.D. An efficient method for the estimation of parameters of stochastic gamma process from noisy degradation measurements. *Proc. Inst. Mech. Eng. Part J. Risk Reliab.* **2013**, *227*, 425–433. [CrossRef]
8. Pulcini, G. A perturbed gamma process with statistically dependent measurement errors. *Reliab. Eng. Syst. Saf.* **2016**, *152*, 296–306. [CrossRef]
9. Giorgio, M.; Mele, A.; Pulcini, G. A perturbed gamma degradation process with degradation dependent non-Gaussian measurement errors. *Appl. Stoch. Model. Bus. Ind.* **2018**, *35*, 198–210. [CrossRef]
10. Meister, A. Density estimation with normal measurement error with unknown variance. *Stat. Sin.* **2006**, *16*, 195–211.
11. Peng, C.Y.; Tseng, S.T. Mis-specification analysis of linear degradation models. *IEEE Trans. Reliab.* **2009**, *58*, 444–455. [CrossRef]

12. Xie, Y.; Wang, X.; Story, M. Statistical methods of background correction for Illumina BeadArray data. *Bioinformatics* **2009**, *25*, 751–757. [CrossRef] [PubMed]
13. Kallen, M.J.; van Noortwijk, J.M. Optimal maintenance decisions under imperfect inspection. *Reliab. Eng. Syst. Saf.* **2005**, *90*, 177–185. [CrossRef]
14. Shen, Y.; Shen, L.; Xu, W. A Wiener-based degradation model with logistic distributed measurement errors and remaining useful life estimation. *Qual. Reliab. Eng. Int.* **2018**, *34*, 1289–1303. [CrossRef]
15. Wang, P.; Tang, Y.; Bae, S.J.; Xu, A. Bayesian Approach for Two-Phase Degradation Data Based on Change-Point Wiener Process With Measurement Errors. *IEEE Trans. Reliab.* **2018**, *67*, 688–700. [CrossRef]
16. Pan, D.; Lu, S.; Liu, Y.; Yang, W.; Liu, J.B. Degradation Data Analysis Using a Wiener Degradation Model With Three-Source Uncertainties. *IEEE Access* **2019**, *7*, 37896–37907. [CrossRef]
17. Sun, H.; Pan, J.; Zhang, J.; Cao, D. Non-linear Wiener process–based cutting tool remaining useful life prediction considering measurement variability. *Int. J. Adv. Manuf. Technol.* **2020**, *107*, 4493–4502. [CrossRef]
18. Tang, S.; Yu, C.; Sun, X.; Fan, H.; Si, X. A Note on Parameters Estimation for Nonlinear Wiener Processes With Measurement Errors. *IEEE Access* **2019**, *7*, 176756–176766. [CrossRef]
19. Liu, D.; Wang, S. Reliability estimation from lifetime testing data and degradation testing data with measurement error based on evidential variable and Wiener process. *Reliab. Eng. Syst. Saf.* **2021**, *205*, 107231. [CrossRef]
20. Li, J.; Wang, Z.; Liu, C.; Qiu, M. Stochastic accelerated degradation model involving multiple accelerating variables by considering measurement error. *J. Mech. Sci. Technol.* **2019**, *33*, 5425–5435. [CrossRef]
21. Sun, B.; Li, Y.; Wang, Z.; Ren, Y.; Feng, Q.; Yang, D. An improved inverse Gaussian process with random effects and measurement errors for RUL prediction of hydraulic piston pump. *Measurement* **2021**, *173*, 108604. [CrossRef]
22. Chen, X.; Ji, G.; Sun, X.; Li, Z. Inverse Gaussian–based model with measurement errors for degradation analysis. *Proc. Inst. Mech. Eng. Part O J. Risk Reliab.* **2019**, *233*, 1086–1098. [CrossRef]
23. Hao, S.; Yang, J.; Berenguer, C. Degradation analysis based on an extended inverse Gaussian process model with skew-normal random effects and measurement errors. *Reliab. Eng. Syst. Saf.* **2019**, *189*, 261–270. [CrossRef]
24. Liu, X.; Wu, Z.; Cui, D.; Guo, B.; Zhang, L. A Modeling Method of Stochastic Parameters' Inverse Gauss Process Considering Measurement Error under Accelerated Degradation Test. *Math. Probl. Eng.* **2019**, *2019*, 1–11. [CrossRef]
25. Chen, X.; Sun, X.; Si, X.; Li, G. Remaining Useful Life Prediction Based on an Adaptive Inverse Gaussian Degradation Process With Measurement Errors. *IEEE Access* **2020**, *8*, 3498–3510. [CrossRef]
26. Plancade, S.; Rozenholc, Y.; Lund, E. Generalization of the normal-exponential model: Exploration of a more accurate parametrisation for the signal distribution on Illumina BeadArrays. *BMC Bioinform.* **2012**, *13*, 329. [CrossRef] [PubMed]
27. Sarder, P.; Nehorai, A. Deconvolution methods for 3-D fluorescence microscopy images. *IEEE Signal Process. Mag.* **2006**, *23*, 32–45. [CrossRef]
28. Swedlow, J.R. Quantitative fluorescence microscopy and image deconvolution. *Methods Cell Biol.* **2013**, *114*, 407–426. [CrossRef]
29. Xu, L.; Ren, J.S.; Liu, C.; Jia, J. Deep convolutional neural network for image deconvolution. *Adv. Neural Inf. Process. Syst.* **2014**, *27*, 1790–1798.
30. Rodriguez-Picon, L.A.; Perez-Dominguez, L.; Mejia, J.; Perez-Olguin, I.J.; Rodriguez-Borbon, M.I. A Deconvolution Approach for Degradation Modeling With Measurement Error. *IEEE Access* **2019**, *7*, 143899–143911. [CrossRef]
31. Zinde-Walsh, V. Measurement error and deconvolution in spaces of generalized functions. *Econom. Theory* **2014**, *30*, 1207–1246. [CrossRef]
32. Wang, X.F.; Wang, B. Deconvolution estimation in measurement error models: The R package decon. *J. Stat. Softw.* **2011**, *39*. [CrossRef]
33. Neumann, M.H. Deconvolution from panel data with unknown error distribution. *J. Multivar. Anal.* **2007**, *98*, 1955–1968. [CrossRef]
34. Lu, C.J.; Meeker, W.O. Using degradation measures to estimate a time-to-failure distribution. *Technometrics* **1993**, *35*, 161–174. [CrossRef]
35. Zhang, M.; Gaudoin, O.; Xie, M. Degradation-based maintenance decision using stochastic filtering for systems under imperfect maintenance. *Eur. J. Oper. Res.* **2015**, *245*, 531–541. [CrossRef]
36. Xie, M.; Goh, T.; Ranjan, P. Some effective control chart procedures for reliability monitoring. *Reliab. Eng. Syst. Saf.* **2002**, *77*, 143–150. [CrossRef]
37. Soltan, H. Advances in Control Charts for Reliability. In Proceedings of the 2019 Industrial & Systems Engineering Conference (ISEC), Jeddah, Saudi Arabia, 19–20 January 2019. [CrossRef]
38. Aslam, M.; Khan, N.; Albassam, M. Control Chart for Failure-Censored Reliability Tests under Uncertainty Environment. *Symmetry* **2018**, *10*, 690. [CrossRef]
39. Faraz, A.; Saniga, E.M.; Heuchenne, C. Shewhart Control Charts for Monitoring Reliability with Weibull Lifetimes. *Qual. Reliab. Eng. Int.* **2014**, *31*, 1565–1574. [CrossRef]
40. Singpurwalla, N.D. Survival in Dynamic Environments. *Stat. Sci.* **1995**, *10*, 86–103. [CrossRef]
41. Van Noortwijk, J. A survey of the application of gamma processes in maintenance. *Reliab. Eng. Syst. Saf.* **2009**, *94*, 2–21. [CrossRef]

42. Bagdonavicius, V.; Nikulin, M. *Accelerated Life Models: Modeling and Statistical Analysis*; Chapman and Hall/CRC Press: Boca Raton, FL, USA, 2001.
43. Bagdonavicius, V.; Nikulin, M.S. Estimation in degradation models with explanatory variables. *Lifetime Data Anal.* **2001**, *7*, 85–103. [CrossRef] [PubMed]
44. Park, C.; Padgett, W. Accelerated degradation models for failure based on geometric Brownian motion and gamma processes. *Lifetime Data Anal.* **2005**, *11*, 511–527. [CrossRef]
45. Park, C.; Padgett, W.J. Stochastic degradation models with several accelerating variables. *IEEE Trans. Reliab.* **2006**, *55*, 379–390. [CrossRef]
46. Brigham, E.O.; Brigham, E.O. *The Fast Fourier Transform*; Prentice-Hall: Englewood Cliffs, NJ, USA, 1974; Volume 7.
47. Cooley, J.W.; Tukey, J.W. An algorithm for the machine calculation of complex Fourier series. *Math. Comput.* **1965**, *19*, 297–301. [CrossRef]
48. Winograd, S. On computing the discrete Fourier transform. *Math. Comput.* **1978**, *32*, 175–199. [CrossRef]
49. Cooley, J.; Lewis, P.; Welch, P. Application of the fast Fourier transform to computation of Fourier integrals, Fourier series, and convolution integrals. *IEEE Trans. Audio Electroacoust.* **1967**, *15*, 79–84. [CrossRef]
50. Abate, J.; Whitt, W. The Fourier-series method for inverting transforms of probability distributions. *Queueing Syst.* **1992**, *10*, 5–87. [CrossRef]
51. Plancade, S.; Whitt, Y. NormalGamma: Normal-Gamma Convolution Model. R Package Version 1.1. 2013. Available online: https://CRAN.R-project.org/package=NormalGamma (accessed on 15 January 2020)
52. Rodríguez-Picón, L.A.; Rodríguez-Picón, A.P.; Méndez-González, L.C.; Rodríguez-Borbón, M.I.; Alvarado-Iniesta, A. Degradation modeling based on gamma process models with random effects. *Commun. Stat.-Simul. Comput.* **2017**, *47*. [CrossRef]
53. Lunn, D.; Spiegelhalter, D.; Thomas, A.; Best, N. The BUGS project: Evolution, critique and future directions. *Stat. Med.* **2009**, *28*, 3049–3067. [CrossRef] [PubMed]
54. Brooks, S.P.; Gelman, A. General methods for monitoring convergence of iterative simulations. *J. Comput. Graph. Stat.* **1998**, *7*, 434–455. [CrossRef]
55. Leiva, V. *The Birnbaum-Saunders Distribution*, 1st ed.; Academic Press: New York, NY, USA, 2016.
56. The Atomotive Industries Action Group. *Measurement Systems Analysis-Reference Manual*; The Atomotive Industries Action Group: Troy, MI, USA, 2002.

Article

Root Cause Analysis of a Printed Circuit Board (PCB) Failure in a Public Transport Communication System

Cher-Ming Tan [1,2,3,4,5], Hsiao-Hi Chen [1], Jing-Ping Wu [1], Vivek Sangwan [1,*], Kun-Yen Tsai [1] and Wen-Chun Huang [1]

1. Center for Reliability Sciences and Technologies, Chang Gung University, Taoyuan 33302, Taiwan; cmtan@cgu.edu.tw (C.-M.T.); emilychen1129@gmail.com (H.-H.C.); cpwu@alumni.nctu.edu.tw (J.-P.W.); tsai.kunyen.hw@gmail.com (K.-Y.T.); weichunhuang2020@gmail.com (W.-C.H.)
2. Department of Electronic Engineering, Chang Gung University, Taoyuan 33302, Taiwan
3. Institute of Radiation Research, Chang Gung University, Taoyuan 33302, Taiwan
4. Center for Reliability Engineering, Ming Chi University of Technology, New Taipei City 24301, Taiwan
5. Department of Radiation Oncology, Chang Gung Memorial Hospital, Taoyuan 33302, Taiwan
* Correspondence: sangwanvivek81@gmail.com

Abstract: A printed circuit board (PCB) is an essential element for practical circuit applications and its failure can inflict large financial costs and even safety concerns, especially if the PCB failure occurs prematurely and unexpectedly. Understanding the failure modes and even the failure mechanisms of a PCB failure are not sufficient to ensure the same failure will not occur again in subsequent operations with different batches of PCBs. The identification of the root cause is crucial to prevent the reoccurrence of the same failure. In this work, a step-by-step approach from customer returned and inventory reproduced boards to the root cause identification is described for an actual industry case where the failure is a PCB burn-out. The failure mechanism is found to be a conductive anodic filament (CAF) even though the PCB is CAF-resistant. The root cause is due to PCB de-penalization. A reliability verification to assure the effectiveness of the corrective action according to the identified root cause is shown to complete the case study. This work shows that a CAF-resistant PCB does not necessarily guarantee no CAF and PCB processes can render its CAF resistance ineffective.

Keywords: 3D X-ray; bias temperature-humidity reliability test; conductive anodic filament (CAF); de-penalization; finite element analysis

1. Introduction

Equipment and devices can fail during their operation in the field and this may occur prematurely. Such a premature failure can render inconveniences and even safety concerns because the failures are unexpected. To prevent the reoccurrence of the failures, a root cause analysis is important. There are differences between failure modes, failure mechanisms, and root causes. An example is an open circuit (failure mode) caused by conductor corrosion (failure mechanism) due to an incomplete protective coating (root cause). Whilst the identification of failure modes can be easy because they can be seen or measured, the finding of failure mechanisms and subsequent root causes are not trivial.

Printed circuit boards (PCBs) are essential for all electronics. They provide mechanical support and electrical connections to the electronic components of an electronic system. As electronic controls and the internet become essential, many public transport systems contain several PCBs. However, the operating environment in public transport can be harsh due to cyclic temperature and humidity, especially in regions with four different weather seasons.

PCB failures range from circuit malfunctions to propagating PCB faults. PCB failures that may appear similar can originate from many different root causes. A propagating circuit board fault is usually considered universally to be a high severity thermal event.

Its initiating mechanism starts when a resistive path forms between two traces or planes in a PCB that are at a different electric potential. This resistive path can form due to an external heat source [1], insulation breakdown [2], arcing [3], or contamination [4]. The power dissipated at the fault region should be sufficient to generate enough heat to further damage the PCB and sustain the thermal event [1]. Slee et al. [1] described various causes of propagating faults in PCBs; namely, resistive heating, interconnect overheating, contamination, electrochemical migration, tin whiskers, insulation failure, and component failure.

In this work, a propagating fault reported was due to electrochemical migration. We report our steps of tracing from the failure modes to the root cause of a PCB used in public transport for an ethernet connection. After their operation for 1.5 months, the operations of the ethernet device stopped and the burn-out area at the bottom edge of the PCB was found, as shown in Figure 1.

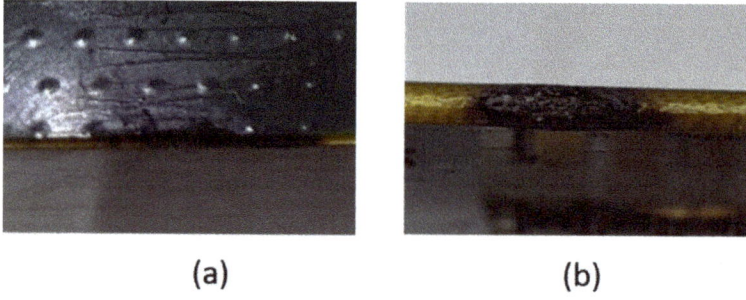

Figure 1. Visual inspection of failed PCBs: (**a**) top view; (**b**) side view.

These PCBs have multilayers that consist of 8 layers with 2 mm of total thickness. Layer 4 is the electrical ground plane and layer 5 is the power plane (V_{IN}) where 12 V is connected to it. Its base material is FR4 (flame retardant). This PCB had superior CAF (conductive anodic filament)-resistance (anti-migration) properties and a high glass transition temperature (T_g) (minimum 170 °C), which makes it a better candidate in the comparison of a standard FR4 for applications where high mechanical and chemical resistances to heat and moisture properties are required.

2. Identification of the Failure Modes

A PCB was returned from customers due to burn-out, as shown in Figure 1. Upon detailed measurements, it was found that the impedance between V_{IN} and GND significantly dropped. A similar burn-out as in Figure 1 was observed on a few originally good inventory boards after testing them with the system in the factory. The supply current was monitored to avoid severe damage to the testing PCBs. It was found that, after the initial short-circuited event, the burn-out expanded due to the over current protection (OCP) mechanisms in the power module. These OCPs are for the protection of the power supply. Two mechanisms are employed for the OCP; namely, the cycle-by-cycle current limit and the cycle skip.

The cycle-by-cycle current limit turns off the power if the maximum current within each pulse width modulation (PWM) cycle is exceeded. The company set the maximum current threshold to 8 A as the continued working current. However, the actual maximum current could be larger than this value due to the response delay; the magnitude depends on the load condition.

The cycle skip mechanism is activated by the continuous occurrence of a cycle-by-cycle current limit condition. The PWM enters into a silent period then restarts the powering cycle again. The company set the continuous occurrence time and the silent period to 30 ms and 250 ms, respectively.

These power supply protection mechanisms can continuously deliver 8.5 W under a short-circuit condition.

3. Identification of the Failure Mechanism

To understand the failure mechanism, a detailed failure analysis was performed. To begin, the structure of the PCB near the burn-out area was studied, as shown in Figure 2. A V-groove penalization was used here; it can protect the PCBs from shocks and vibration experienced during paste printing to component assembly, soldering, and even testing [5].

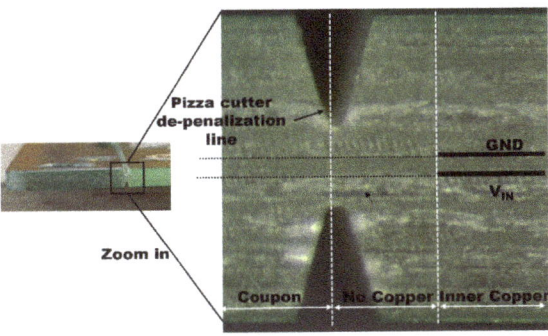

Figure 2. Cross-section of the board near the V-cut scoring area.

After the PCB fabrication processes are completed, de-penalization is performed to split them into individual PCBs. Currently, there are various types of de-penalization; namely, hand break, pizza cutter/V-cut, punch, router, saw, and laser [5]. In this work, a pizza cutter/V-cut was employed for the de-penalization.

Upon careful examination of the PCBs that were not short-circuited or burned out, the presence of many crack lines could be seen (Figure 3). These crack lines were found after the de-penalization. With these crack lines, moisture from the environment could penetrate and diffuse into the PCB easily. In the presence of the applied voltage, the electric field could cause the copper ions from the PCB copper planes to drift across either as a dendrite formation [1,5–8] or as a CAF formation [9,10]. This was believed to be the failure mechanism.

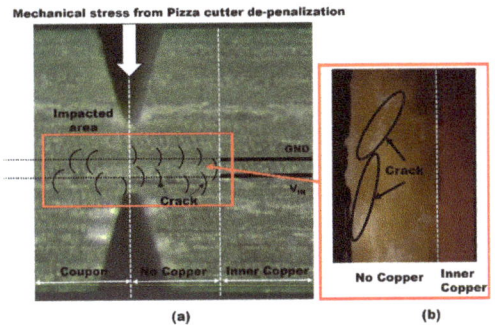

Figure 3. Cross-section of the PCB: (**a**) potential crack location; (**b**) after pizza cutter (V-cut) de-panel process.

Dendritic growth usually occurs when metal ions move into the electrolyte layer near the anode and then deposit near the cathode and grow in a tree-like or needle-like formation. IPC-5704 defines dendritic growth as the growth of conductive metal filaments on a PCB through an electrolytic solution under the influence of a DC voltage bias and it is usually formed on the surface of a PCB [11]. Heat can be generated from the electrical ionic conduction current during the migration process or from the electrical current conducted through the dendrites when they bridge the conductors. Eventually, the dendrites can develop into substantially low resistance with significant heat dissipation and can result in a propagating fault in a PCB [1].

The CAFs are formed as anode metal gradually becomes ions and begins to migrate under the effects of the bias voltage. It is formed within the PCB insulation materials. Unlike dendrite formation, the conductive filament growth forms from the anode to the cathode, hence the name conductive anodic filament [1]. Failure symptoms of a CAF range from intermittent resistive faults under current limit conditions to catastrophic propagating circuit board faults if the fault current is not sufficiently limited [1]. Basically, CAF formation follows a two-step process; namely, the formation of a microcrack that bridges two conductors to establish the migration path and an electrochemical reaction with the presence of a DC bias and moisture to drive the migration along the cracks between the two conductors [12]. This was observed in our experiments, as will be shown later.

The formation of dendrites and CAFs can seriously influence the reliability of electronic systems. From our failure analysis observations, the failure mechanism was likely to be a CAF as the short-circuit paths were not on the surface although this PCB was CAF-resistant.

4. Mechanism of the Copper Particles in the FR4 during CAF

Due to the presence of moisture and an electric field, copper ions will migrate along the glass fiber. This is a well-established phenomenon in PCBs known as CAF. As copper ions move, and as it is known that conductive particles exposed to an applied electric field will be charged [13,14], these copper ions will also be charged. As the charged particles move, they also induce a current through the external circuit [15] and, in the case of partial discharge (which will be described later), a very short pulse of current will be induced as there will be a charge transfer between the particle and the facing electrode, which is the cathode [16].

Generally, conductive particles in the presence of an electric field and fluid such as water will be subjected to various forces including a drift electric force, fluid force, and gravity force. The fluid force includes a mass force, drag force, and Basset force. For a detailed study of the forces, one may refer to the work by Pan et al. [17]. Although the work by Pan is for conductive particles in liquids, several of the underlying physical mechanisms considered can be applied. In the situation studied in this work, due to its mass and density, the copper particles were only subjected to the drift electric force and no rebound between the two electrodes occurred.

As the conductive particles are close to the cathode as driven by the electric field from the anode to the cathode, partial discharge can occur due to the intensification of the electric field between them [17]. It has been found that the electric field is enhanced by 4.2 times for a spherical particle [16]. In the presence of multiple conductive particles, as in our case, the frequency of the partial discharge can be much higher, as shown by Li et al. [18].

However, partial discharge does not necessary lead to the breakdown of the dielectric but it will lead to the reduction of its breakdown voltage. As the size and concentration of the particles increase, the insulating performance of the dielectric is further inhibited [19–21]. With the reduction in the insulating performance, a leakage current can flow through the FR4 and generate heat. It is worth noting that if the temperature around the FR4 exceeds more than 65 °C, heat dissipation becomes a major issue for the FR4; beyond this temperature, the failure mechanism changes to a thermal breakdown from an electric breakdown [2].

When the particle concentration is relatively high, the formation of a particle bridge is possible. It was found that a DC field is beneficial to such a formation, as shown by the work by Li et al. [22].

After a partial discharge, it is suspected that a short carbonization path is formed that connects the conductive particles to the nearby electrode [2,23]. In a sense, the electrode is now extruded. Hence, the conductive particles that are further away can now also have a partial discharge with the subsequent formation of a carbonization path. When the resistance reduces to a certain value that corresponds with an increase in the leakage current, the Joule heating can drive out the surrounding moisture and an increase in resistance can be expected, as was observed in our experiments (shown later in Section 7) However, the charge relaxation on the conductive particles may also be reduced and thus the movement of the charged conductive particles can move further. At the same time, this localized heat enlarges the crack. Hence, external moisture can now diffuse further and in a greater amount into the FR4. As the conductive particles move closer to the extruded electrode, partial discharge occurs again. This iteration continues until the effective distance between the extruded electrode and the anode becomes so close that the field between them is so high, it exceeds the breakdown field of the FR4, which is already reduced due to the presence of conductive particles and carbonization.

5. Experimental Results and Discussion

To confirm the proposed mechanism, a Thermo Scientific ELITE VX system was used to identify the hot spot of the thermal emissions and a 3D X-ray was performed using ZEISS Xradia 520 Versa on the hot spot. Figure 4 shows the enlarged X-ray micrograph at the hot spot in the burn-out area and we can clearly see that the short-circuit was due to copper particles at the edge of the PCB along the de-penalized line. From the cross-sectional view of the X-ray micrograph, one can see that the trace of copper is indeed from the V_{IN} (anode) toward the GND (ground).

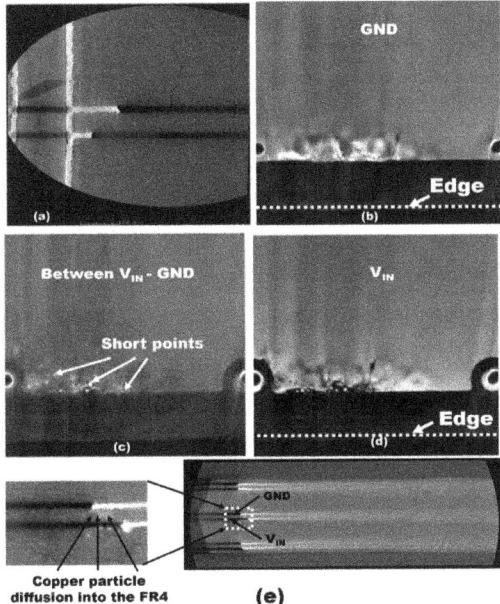

Figure 4. A 3D X-ray of the PCB that shows the cross-section of the PCB. The propagation of the cracks and the movement of copper from the V_{IN} layer to the GND layer due to the presence of the water layer and electric field can be seen: (**a**) shows the formation of crack lines before the resistance between them begins to decrease; (**b**,**d**) are the plane of the GND and V_{IN} layer, respectively, as extracted from the 3D X-ray; (**c**) is the plane in the middle of the GND layer and V_{IN} layer as extracted from the 3D X-ray where the white arrows represent copper particles; (**e**) side view of the PCB showing the short point in between the V_{IN} layer and GND layer.

To further analyze and characterize the cracks formed in the PCB, as shown in Figure 4, the PCB area with the cracks was dissected out and examined under an X-ray Micro CT (computed tomography) scanner system (model number CTLab-HX130-2-E). The dissection was needed due to the size limitation of the X-ray system. For a better view of the cracks, the image dataset was reoriented with a new image plane, as shown by the pink color plane in the top left side image in Figure 5. In Figure 5 (top left), 'a', 'b', and 'c' are marked to show the front, top, and left sides, respectively, and their respective 2D images are shown in the same figure.

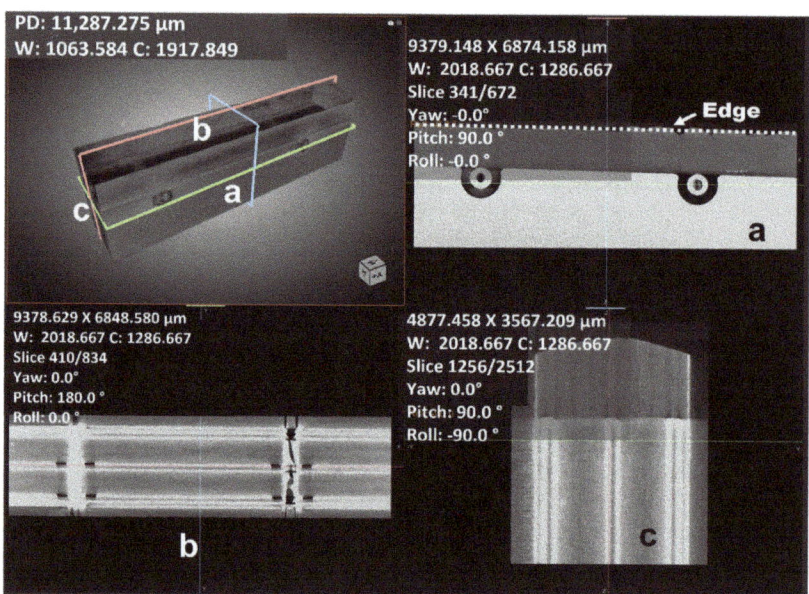

Figure 5. Image dataset reoriented with a new image plane as shown by the pink color in the 3D image (top left side). The 2D images are taken at the: front side (**a**); top side (**b**); left side (**c**).

A total of 78 cracks were identified, and these cracks are shown by their individual colors in Figure 6. The histogram of the volume of these cracks is shown in Figure 7a, which shows that the majority of the cracks (25) had a volume less than 1×10^5 (μm^3). No crack had a volume greater than 4×10^5 (μm^3). Figure 7b shows all the identified cracks and the cracks with the same volume are shown with the same colors in the 3D view. It could be observed that large cracks occurred at the edge of the PCB and they progressed toward the inner parts of the PCB along with the GND and V_{IN} layers. All these cracks were due to excessive high strain at a high strain rate induced during the V-cut de-panelization as measured experimentally with the results shown in Figure 8. The experimentation is discussed in the next section.

The large cracks observed were concentrated in a certain area of the PCB instead of being distributed uniformly. This non-uniform distribution of the crack sizes was believed to be due to the inhomogeneous strain distribution along the edge of the PCB during the V-cut de-panelization as simulated using a finite element analysis, which will be discussed in the next section. There were also large cracks present away from the edge but they were closed to the largest cracks, indicating the weakening of the mechanical integrity of the board after the large cracks occurred.

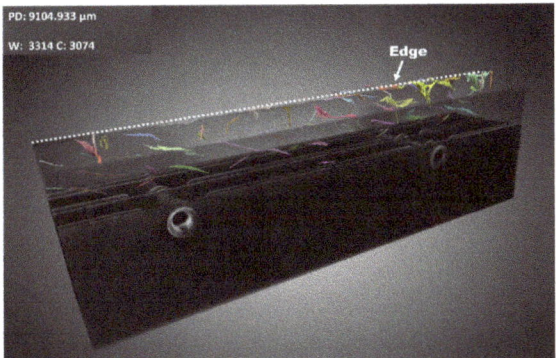

Figure 6. A total of 78 cracks are identified and represented by individual colors.

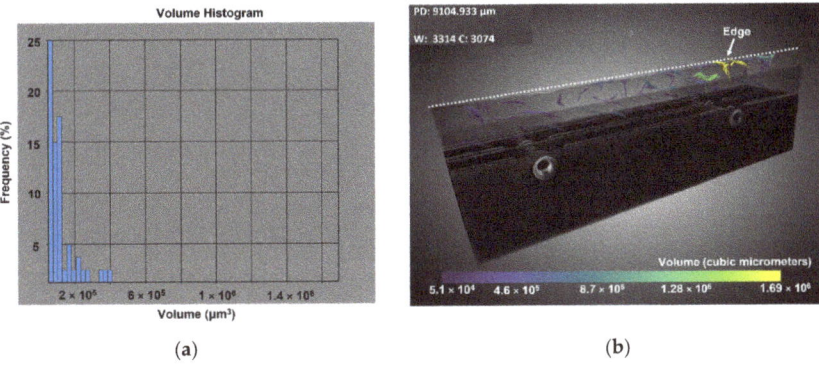

(a)   (b)

Figure 7. (a) Volume histogram of the cracks shown in Figure 6; (b) cracks with the same volume are shown with the same color.

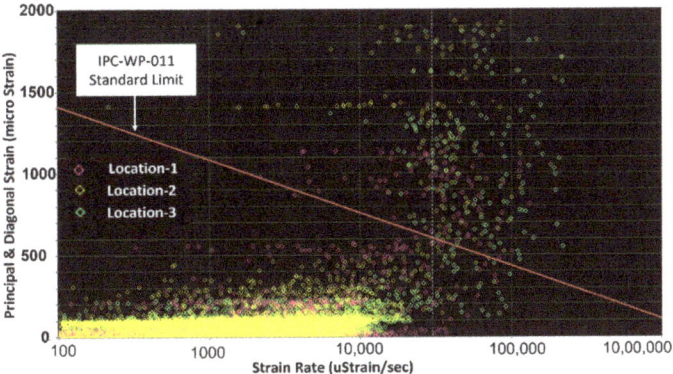

Figure 8. Strain gauge measurement data show that a part of the principal strain exceeds the IPC-WP-011 standard limits during the cutting process.

To further confirm the proposed mechanism, 5 samples of the PCB that had crack lines due to de-penalization were subjected to 65 °C/95% RH. Table 1 summarizes the test results. The test condition was selected so that a continuous film of water could be formed in the PCB. The test results showed clearly that these cracks allowed moisture to diffuse into the PCB and reduced the impedance between the GND and V_{IN}.

Table 1. Change in resistances after temperature/humidity test.

Sample #	Initial Resistance between V_{IN} and GND (Ω)	Measured Resistance after the Specified Test Duration (Ω)	Test Duration (Hours)
1	Over 10^8	58.6	160
2	Over 10^8	30.9 K	113
3	Over 10^8	316.2	90
4	Over 10^8	1.63 K	232
5	Over 10^8	46.5	141

From the above observations, the failure mechanism could be summarized, as shown in Figure 9. The flow chart and its summary table to represent the research methodology of this work are shown in Figure 10 and Table 2, respectively.

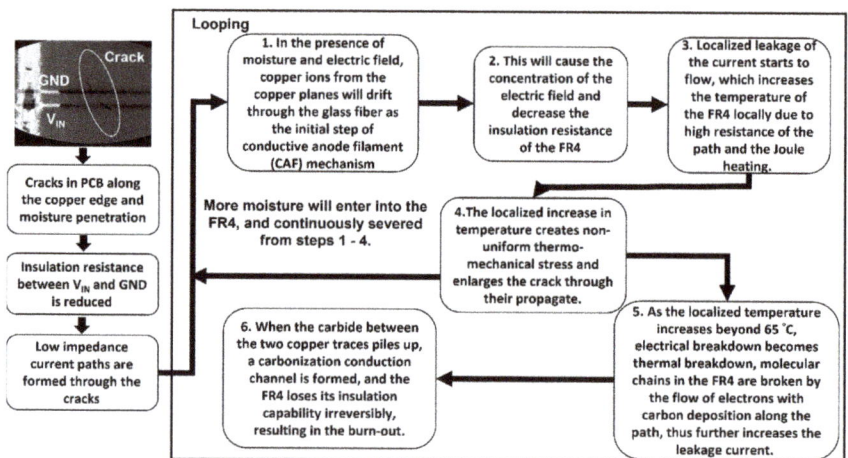

Figure 9. Failure process and the underlying mechanisms.

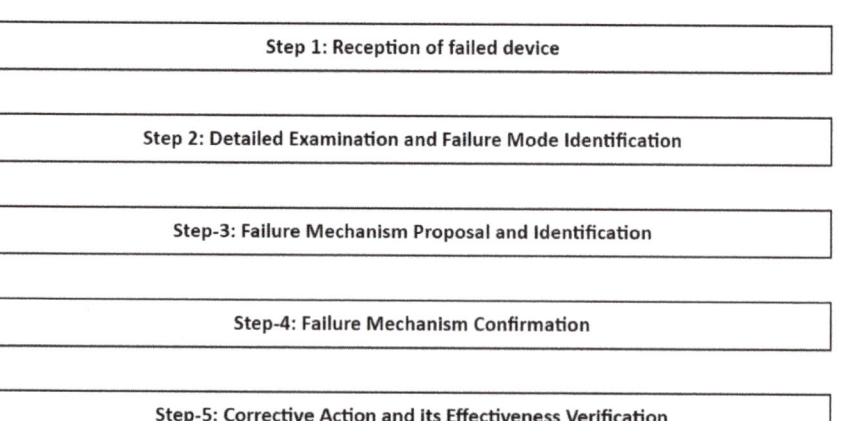

Figure 10. Flow chart of the failure analysis process.

Table 2. Summary of the failure analysis methodology.

Steps	Failure Analysis Methodology	Details and Findings
1	Reception of Failed Device	Customer reported unavailability of ethernet
2	Detailed Examination and Failure Mode Inspection	• Burn-out area at the bottom edge of the PCB found through naked eyes. • Presence of cracks observed at the burned-out area using a low- to high-definition optical microscope, thermal emissions, 3D X-ray, and X-ray Micro CT Scanning. • Cracks are found in between the L4~L5 layers at the PCB edge.
3	Failure Mechanism Proposal and Identification	• Cracks are present at the PCB edge from where the moisture penetrates inside the PCB, which causes the short-circuit; the burned-out area can be observed. • Conductive anodic filament (CAF) is a proposed failure mechanism. • Strain produced during the V-cut de-panelization process is more than IPC-WP-011 standard limits and verified using a strain gauge measurement and an ANSYS multiphysics simulation, which are the causes behind the cracks in the PCB.
4	Failure Mechanism Confirmation	PCBs are subjected to a 65 °C/95% RH reliability test and resistance degradation is observed, which signifies that the moisture penetrates into the PCBs through the cracks that are present in between the L4~L5 layers.
5	Corrective Action and its Effectiveness Verification	• New de-panelization process is introduced. • PCBs are prepared using the new de-panelization process subjected to 65 °C/95% RH. • No cracks and degradation in the PCB resistance are observed

6. Possible Root Causes and Their Verification

Knowing that it was the crack lines that caused the observed failure mode, and that these lines were generated because of the high strain developed during the de-penalization process, a strain gauge measurement was performed on the PCB at three test locations, as marked by the black color blocks in Figure 11, near the edge of the PCB during the V-cutting process to ascertain the root cause. The strain gauge measurement results are depicted in Figure 8. One can see the localized areas along the de-paneling line at the bottom side with a high strain rate and high strain around 1400–2050, which was beyond the acceptable criteria according to IPC-WP-011 [24].

ANSYS WorkBench 19.0 was also employed to examine the strain induced by de-penalization. A 3D CAD model of the PCB was designed in ANSYS WorkBench and a V-cut cutter was designed to mimic the real-world scenario in the simulation software where mechanical strain was observed, as shown in Figure 11. Our simulation results showed the maximum stress area (red area) distributed in strips along the edge of the PCB and this coincided with the burn-out area. The maximum strains did not distribute uniformly, which agreed with our X-ray micrograph and the strain gauge measurement results. Li [25] also reported that the stress generated through the V-cut process is significantly high in comparison with other cutting processes that cause mechanical cracks in PCBs.

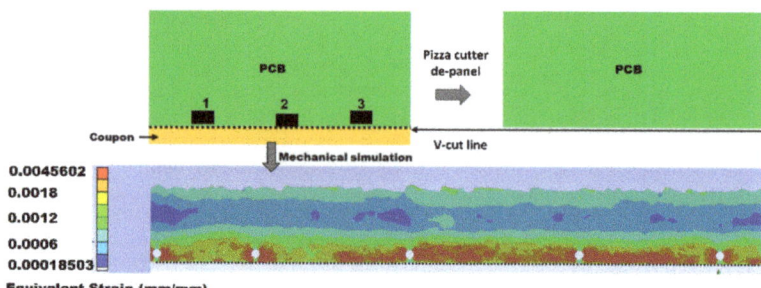

Figure 11. PCB de-panel process and mechanical simulation result. Locations 1, 2, and 3 show strain gauge measurement points.

In the pizza cut de-penalizing, the applied mechanical stress caused cracks on the copper keep-out area of the PCB, as shown in Figure 12. However, without de-paneling, there was no crack on the copper keep-out area, as can be seen in Figure 13.

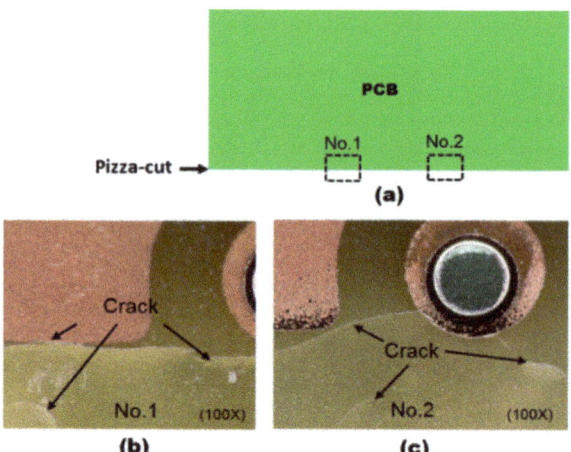

Figure 12. Observation of cracks in a PCB after de-paneling using an Olympus BX51 microscope at 50 × magnification of: (**b**) location No. 1 and (**c**) location No. 2, as marked in (**a**), which is a horizontal cross-section of a PCB with a pizza cutter de-panel.

It was, therefore, clear that it was the pizza cutter de-penalization that caused the cracks in the PCB, which then resulted in moisture diffusion into the PCB. As the cutting area was near the GND plane and V_{IN} plane, as can be seen in Figure 2, the reported failure mode was observed. We found that no damage was observed if a manual de-penalization was performed. Thus, we believed that a proper V-groove design and machine setting could solve this issue, but this was beyond the scope of this work. With proper settings such as the incident angle, rotary speed, and cooling fluid, such damage can be minimized, as in the case of wafer dicing [26].

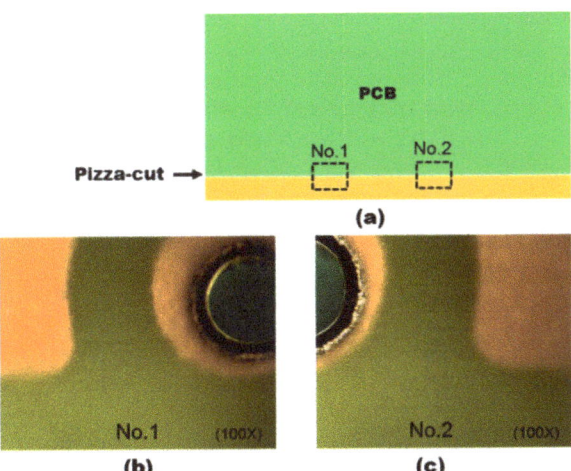

Figure 13. Absence of cracks in a PCB without de-paneling. Observations were performed using an Olympus BX51 microscope at 50× magnification of: (**b**) location No. 1 and (**c**) location No. 2, as marked in (**a**), which is a horizontal cross-section location of a PCB without a de-panel.

7. Corrective Actions and Reliability Verification

With the identification of the root cause, another type of de-penalization was introduced where no crack was produced. Due to confidential considerations, this new way cannot be disclosed.

To verify the effectiveness of the corrective actions, reliability tests were performed along with the new de-penalization method. The test condition was set at 65 °C/95% RH so that a continuous water film could be formed if cracks existed. To speed up the test, a voltage of 100 V was applied across the GND and V_{IN}.

From Figure 14, with the new de-penalization method, the resistance between the GND and V_{IN} remained high: up to 800 h of testing consistently for the 10 test samples. However, 4 PCBs with the V-cut de-penalizing methods showed a significant reduction in the resistance even before 400 h of testing, except for one board. This clearly showed that our corrective action was effective in overcoming the damage introduced due to de-penalization.

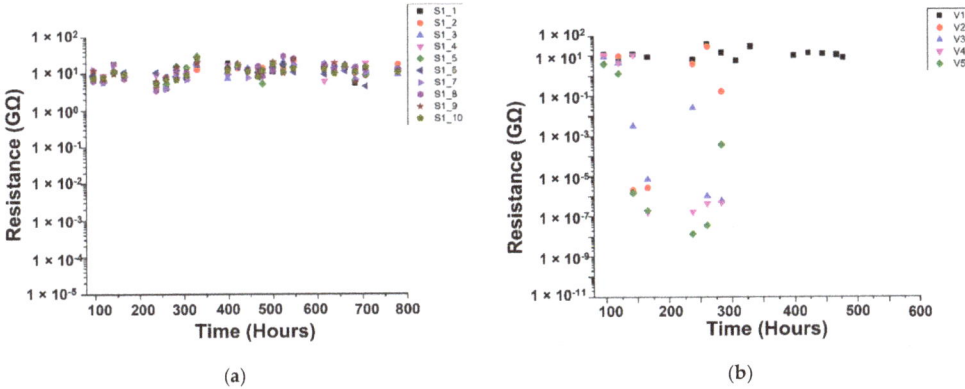

Figure 14. Resistance changes over time under the reliability test mentioned above: (**a**) 10 samples with the new de-penalization; (**b**) 5 samples with pizza cut de-penalization and the resistance fluctuation as reported earlier can be observed. The resistance in the Y-axis is in GW.

8. Conclusions

This work demonstrated a step-by-step failure analysis methodology for multilayer printed circuit boards that led an observed failure mode to the root cause with verification of the root cause and the effectiveness of the corresponding corrective action. Printed circuit boards used in public transport systems were found to be burnt after a time of operation. A detailed failure analysis showed that the failure mechanism was a propagating fault of a CAF formation as moisture was diffused into the PCBs via microcracks introduced due to a V-cut de-penalization. Although the PCB itself was CAF-resistant, the presence of a microcrack due to de-penalization could render its CAF resistance ineffective.

With the identified root cause, a modified de-penalization method was developed and no microcracks were observed with this new method, which verified the root cause. The PCBs with the modified de-penalization method also underwent reliability tests under a high temperature/humidity with a high voltage applied across and only small changes in resistance between the previously affected planes were observed, which confirmed the effectiveness of the corrective action. The detailed steps from the failure mode to the failure mechanism and from the failure mechanism to the root cause were clearly described.

Author Contributions: Methodology, C.-M.T. and J.-P.W.; formal analysis, C.-M.T., H.-H.C., J.-P.W., and V.S.; investigation, C.-M.T., J.-P.W., and V.S.; resources, H.-H.C., K.-Y.T., and W.-C.H.; writing—original draft preparation, V.S.; writing—review and editing, C.-M.T., J.-P.W., H.-H.C., and V.S.; project administration, H.-H.C. All authors have read and agreed to the published version of the manuscript.

Funding: This research received no external funding.

Acknowledgments: This research is an industrial collaboration with a company to solve a real industrial issue, leveraging on the in-depth applied failure analysis theory developed by the Centre for Reliability Science and Technology in Chang Gung University. This research did not receive any specific grants from funding agencies in the public, commercial, or not-for-profit sectors. The authors also acknowledge the support of C.Y. Hsueh from Jie Dong, Taiwan and Rigaku Corporation, Japan, in assisting with the X-ray computed tomography analysis.

Conflicts of Interest: The authors declare no conflict of interest.

References

1. Slee, D.; Stepan, J.; Wei, W.; Swart, J. Introduction to printed circuit board failures. In Proceedings of the 2009 IEEE Symposium on Product Compliance Engineering, Toronto, ON, Canada, 26–28 October 2009; pp. 1–8.
2. Zhou, Q.; Wen, M.; Xiong, T.; Jiang, T.; Zhou, M.; Ouyang, X.; Xing, L. Study on Insulation Breakdown Characteristics of Printed Circuit Board under Continuous Square Impulse Voltage. *Energies* **2018**, *11*, 2908. [CrossRef]
3. Davis, J.H. Silicone protective encapsulants and coatings for electronic components and circuits. In *Plastics for Electronics*; Springer: Dordrecht, The Netherlands, 1985; pp. 67–97.
4. Khandpur, R.S. *Printed Circuit Boards: Design, Fabrication, Assembly and Testing*; Tata McGraw-Hill Electronic Engineering: New York, NY, USA, 2005; ISBN 0071589252.
5. Coombs, C.F.J.; Holden, H.T. *Printed Circuits Handbook*, 7th ed.; McGraw-Hill Education: New York, NY, USA, 2016.
6. Minzari, D.; Jellesen, M.S.; Møller, P.; Ambat, R. On the electrochemical migration mechanism of tin in electronics. *Corros. Sci.* **2011**, *53*, 3366–3379. [CrossRef]
7. Yang, S.; Wu, J.; Christou, A. Initial stage of silver electrochemical migration degradation. *Microelectron. Reliab.* **2006**, *46*, 1915–1921. [CrossRef]
8. Lee, S.B.; Jung, M.S.; Lee, H.Y.; Kang, T.; Joo, Y.C. Effect of bias voltage on the electrochemical migration behaviors of Sn and Pb. *IEEE Trans. Device Mater. Reliab.* **2009**, *9*, 483–488. [CrossRef]
9. Tan, C.M.; Narula, U.; Seow, G.L.; Sangwan, V.; Chen, C.H.; Lin, S.P.; Chen, J.Y. Moisture resistance evaluation on single electronic package moulding compound. *J. Mater. Chem. C* **2020**, *8*, 1943–1952. [CrossRef]
10. Ready, W.J.; Stock, S.R.; Freeman, G.B.; Dollar, L.L.; Turbini, L.J. Microstructure of Conductive Anodic Filaments Formed during Accelerated Testing of Printed Wiring Boards. *Circuit World* **1995**, *21*, 5–9. [CrossRef]
11. IPC. *IPC-5704 Cleanliness Requirements for Unpopulated Printed Boards*; IPC: Bannockburn, IL, USA, 2009.
12. Zou, S.; Li, X.; Dong, C.; Ding, K.; Xiao, K. Electrochemical migration, whisker formation, and corrosion behavior of printed circuit board under wet H2S environment. *Electrochim. Acta* **2013**, *114*, 363–371. [CrossRef]
13. Pérez, A.T. Charge and force on a conducting sphere between two parallel electrodes. *J. Electrostat.* **2002**, *56*, 199–217. [CrossRef]

14. Drews, A.M.; Kowalik, M.; Bishop, K.J.M. Charge and force on a conductive sphere between two parallel electrodes: A Stokesian dynamics approach. *J. Appl. Phys.* **2014**, *116*, 074903. [CrossRef]
15. Pan, C.; Wu, K.; Du, Y.; Tang, J. Comparison of Sato's equation and Pedersen's theory to obtain gas discharge current. *IEEE Trans. Dielectr. Electr. Insul.* **2016**, *23*, 1690–1698. [CrossRef]
16. Tobazéon, R. Electrohydrodynamic behaviour of single spherical or cylindrical conducting particles in an insulating liquid subjected to a uniform DC field. *J. Phys. D Appl. Phys.* **1996**, *29*, 2595–2608. [CrossRef]
17. Pan, C.; Tang, J.; Chen, G.; Zhang, Y.; Luo, X. Review about PD and breakdown induced by conductive particles in an insulating liquid. *High Volt.* **2020**, *5*, 287–297. [CrossRef]
18. Li, J.; Hu, Q.; Zhao, X.; Yao, X.; Luo, Y.; Li, Y. Partial-discharge characteristics of free spherical conducting particles under AC condition in transformer oils. *IEEE Trans. Power Deliv.* **2011**, *26*, 538–546. [CrossRef]
19. Zhang, J.; Wang, F.; Li, J.; Ran, H.; Huang, D. Influence of copper particles on breakdown voltage and frequency-dependent dielectric property of vegetable insulating oil. *Energies* **2017**, *10*, 938. [CrossRef]
20. Wang, K.; Wang, F.; Li, J.; Zhao, Q.; Wen, G.; Zhang, T. Effect of metal particles on the electrical properties of mineral and natural ester oils. *IEEE Trans. Dielectr. Electr. Insul.* **2018**, *25*, 1621–1627. [CrossRef]
21. Wang, X.; Wang, Z.D. Study of dielectric behavior of ester transformer liquids under ac voltage. *IEEE Trans. Dielectr. Electr. Insul.* **2012**, *19*, 1916–1925. [CrossRef]
22. Li, Y. Motion Characteristics of Solid Particles in Insulating Oil and Their Influence on Breakdown Voltage of Insulating Oil. Master's Thesis, Chongqing University, Chongqing, China, 2017.
23. ECSS. *European Cooperation for Space Standardization Space Engineering: High Voltage Engineering and Design Handbook*; ECSS: Noordwijk, The Netherlands, 2012.
24. IPC. *IPC-WP-011 Guidance for Strain Gage Limits for Printed Circuit Assemblies*; IPC: Bannockburn, IL, USA, 2012.
25. Li, W.; Sun, X. An analysis case on the failure of BGA solder joints. In Proceedings of the 18th International Conference on Electronic Packaging Technology, ICEPT 2017, Harbin, China, 16–19 August 2017; pp. 731–734.
26. Shi, K.W.; Yow, K.Y. The characteristics and factors of a wafer dicing blade and its optimized interactions required for singulating high metal stack lowk wafers. In Proceedings of the 2013 IEEE 15th Electronics Packaging Technology Conference, EPTC 2013, Singapore, 11–13 December 2013; pp. 208–212.

MDPI
St. Alban-Anlage 66
4052 Basel
Switzerland
Tel. +41 61 683 77 34
Fax +41 61 302 89 18
www.mdpi.com

Applied Sciences Editorial Office
E-mail: applsci@mdpi.com
www.mdpi.com/journal/applsci

www.ingramcontent.com/pod-product-compliance
Lightning Source LLC
LaVergne TN
LVHW070641100526
838202LV00013B/853